Lecture Notes in Physics

For information about Vols. 1–23, please contact your bookseller or Springer-Verlag.

Vol. 24: R. F. Snipes, Statistical Mechanical Theory of the Electrolytic Transport of Nonelectrolytes. V, 210 pages. 1973.

Vol. 25: Constructive Quantum Field Theory. The 1973 "Ettore Majorana" International School of Mathematical Physics. Edited by G. Velo and A. Wightman. III, 331 pages. 1973.

Vol. 26: A. Hubert, Theorie der Domänenwände in geordneten Medien. XII, 377 Seiten. 1974.

Vol. 27: R. K. Zeytounian, Notes sur les Ecoulements Rotationnels de Fluides Parfaits. XIII, 407 pages. 1974.

Vol. 28: Lectures in Statistical Physics. Edited by W. C. Schieve and J. S. Turner. V, 342 pages. 1974.

Vol. 29: Foundations of Quantum Mechanics and Ordered Linear Spaces. Advanced Study Institute, Marburg 1973. Edited by A. Hartkämper and H. Neumann. VI, 355 pages. 1974.

Vol. 30: Polarization Nuclear Physics. Proceedings 1973. Edited by D. Fick. IX, 292 pages. 1974.

Vol. 31: Transport Phenomena. Sitges International Schools of Statistical Mechanics, June 1974. Edited by G. Kirczenow and J. Marro. XIV, 517 pages. 1974.

Vol. 32: Particles, Quantum Fields and Statistical Mechanics. Proceedings 1973. Edited by M. Alexanian and A. Zepeda. V, 132 pages. 1975.

Vol. 33: Classical and Quantum Mechanical Aspects of Heavy Ion Collisions. Proceedings 1974. Edited by H. L. Harney, P. Braun-Munzinger, and C. K. Gelbke. VII, 311 pages. 1975.

Vol. 34: One-Dimensional Conductors GPS Summer School Proceedings, 1974. Edited by H. G. Schuster. VII, 371 pages. 1975.

Vol. 35: Proceedings of the Fourth International Conference on Numerical Methods in Fluid Dynamics, 1974. Edited by R. D. Richtmyer. V, 457 pages. 1975.

Vol. 36: R. Gatignol, Théorie Cinétique des Gaz à Répartition Discrète de Vitesses. II, 219 pages. 1975.

Vol. 37: Trends in Elementary Particle Theory. Proceedings 1974. Edited by H. Rollnik and K. Dietz. V, 472 pages. 1975.

Vol. 38: Dynamical Systems, Theory and Applications. Proceedings 1974. Edited by J. Moser. VI, 624 pages. 1975.

Vol. 39: International Symposium on Mathematical Problems in Theoretical Physics. Proceedings 1975. Edited by H. Araki. XII, 562 pages. 1975.

Vol. 40: Effective Interactions and Operators in Nuclei. Proceedings 1975. Edited by B. R. Barrett. XII, 339 pages. 1975.

Vol. 41: Progress in Numerical Fluid Dynamics. Proceedings 1974. Edited by H. J. Wirz. V, 471 pages. 1975.

Vol. 42: H II Regions and Related Topics. Proceedings 1975. Edited by D. Downes and T. L. Wilson. XII, 488 pages. 1975.

Vol. 43: Laser Spectroscopy. Proceedings 1975. Edited by S. Haroche, J. C. Pebay-Peyroula, T. W. Hänsch, and S. E. Harris. X, 466 pages. 1975.

Vol. 44: R. A. Breuer, Gravitational Perturbation Theory and Synchrotron Radiation. VI, 196 pages. 1975.

Vol. 45: Dynamical Concepts on Scaling Violation and the New Resonances in e^+e^- Annihilation. Edited by B. Humpert. VII, 248 pages. 1976.

Vol. 46: E. J. Flaherty, Hermitian and Kählerian Geometry in Relativity. VIII, 365 pages. 1976.

Vol. 47: Padé Approximants Method and Its Applications to Mechanics. Edited by H. Cabannes. XV, 267 pages. 1976.

Vol. 48: Interplanetary Dust and Zodiacal Light. Proceedings 1975. Edited by H. Elsässer and H. Fechtig. XII, 496 pages. 1976.

Vol. 49: W. G. Harter and C. W. Patterson, A Unitary Calculus for Electronic Orbitals. XII, 144 pages. 1976.

Vol. 50: Group Theoretical Methods in Physics. 4th International Colloquium. Nijmegen 1975. Edited by A. Janner, T. Janssen, and M. Boon. XIII, 629 pages. 1976.

Vol. 51: W. Nörenberg und H. A. Weidenmüller. Introduction to the Theory of Heavy-Ion Collisions. IX, 273 pages. 1976.

Vol. 52: M. Mladjenović, Development of Magnetic β-Ray Spectroscopy. X, 282 pages. 1976.

Vol. 53: D. J. Simms and N. M. J. Woodhouse, Lectures on Geometric Quantization. V, 166 pages. 1976.

Vol. 54: Critical Phenomena. Sitges International School on Statistical Mechanics, June 1976. Edited by J. Brey and R. B. Jones. XI, 383 pages. 1976.

Vol. 55: Nuclear Optical Model Potential. Proceedings 1976. Edited by S. Boffi and G. Passatore. VI, 221 pages. 1976.

Vol. 56: Current Induced Reactions. International Summer Institute, Hamburg 1975. Edited by J. G. Körner, G. Kramer, and D. Schildknecht. V, 553 pages. 1976.

Vol. 57: Physics of Highly Excited States in Solids. Proceedings 1975. Edited by M. Ueta and Y. Nishina. IX, 391 pages. 1976.

Vol. 58: Computing Methods in Applied Sciences. Proceedings 1975. Edited by R. Glowinski and J. L. Lions. VIII, 593 pages. 1976.

Vol. 59: Proceedings of the Fifth International Conference on Numerical Methods in Fluid Dynamics. 1976. Edited by A. I. van de Vooren and P. J. Zandbergen. VII, 459 pages. 1976.

Vol. 60: C. Gruber, A. Hintermann, and D. Merlini, Group Analysis of Classical Lattice Systems. XIV, 326 pages. 1977.

Vol. 61: International School on Electro and Photonuclear Reactions I. Edited by C. Schaerf. VIII, 650 pages. 1977.

Vol. 62: International School on Electro and Photonuclear Reactions II. Edited by C. Schaerf. VIII, 301 pages. 1977.

Vol. 63: V. K. Dobrev et al., Harmonic Analysis on the n-Dimensional Lorentz Group and Its Application to Conformal Quantum Field Theory. X, 280 pages. 1977.

Vol. 64: Waves on Water of Variable Depth. Edited by D. G. Provis and R. Radok. 231 pages. 1977.

Vol. 65: Organic Conductors and Semiconductors. Proceedings 1976. Edited by L. Pál, G. Grüner, A. Jánossy and J. Sólyom. 654 pages. 1977.

Vol. 66: A. H. Völkel, Fields, Particles and Currents. VI, 354 pages. 1977.

Vol. 67: W. Drechsler and M. E. Mayer, Fiber Bundle Techniques in Gauge Theories. X, 248 pages. 1977.

Lecture Notes in Physics

Edited by J. Ehlers, München, K. Hepp, Zürich
R. Kippenhahn, München, H. A. Weidenmüller, Heidelberg
and J. Zittartz, Köln
Managing Editor: W. Beiglböck, Heidelberg

111

H.-J. Schmidt

Axiomatic Characterization of Physical Geometry

Springer-Verlag
Berlin Heidelberg GmbH 1979

Author

Heinz-Jürgen Schmidt
Fachbereich 5
Naturwissenschaften/Mathematik
Universität Osnabrück
Postfach 4469
D-4500 Osnabrück

ISBN 978-3-540-09719-8

Library of Congress Cataloging in Publication Data
Schmidt, Heinz-Jürgen, 1948-
Axiomatic characterization of physical geometry.
(Lecture notes in physics; 111)
Bibliography: p.
Includes index.
1. Geometry. 2. Axiomatic set theory. I. Title. II. Series.
QC20.7.G44S35 530.1'5162 79-23944
ISBN 978-3-540-09719-8 ISBN 978-3-540-38515-8 (eBook)
DOI 10.1007/978-3-540-38515-8

Originally published by Springer-Verlag Berlin Heidelberg New York in 1979

2153/3140-543210

PREFACE

This book will deal with the basis of a theory, which can be considered as the most ancient part of physics, namely Euclidean geometry. For about 100 years there has been a debate on the physical space problem, especially stimulated by the creation of (non-Euclidean) General Relativity. In spite of this, contrary to the impression generated by some textbooks on physics, the topic is far from being in a final form. The problems of interpretations and definitions of physical concepts are often neglected, partly because methodological rigor is (successfully) replaced by physical intuition, and partly because these problems are inherently difficult and inextricably intertwined. In contrast to the situation in mathematics, the foundations of physics are still in their pre-Bourbaki millenium. I think, however, G. Ludwig has made an important step toward an adequate understanding of physics, and this book may be viewed as a partial realization of one point of his program.

A large class of physical applications of Euclidean geometry concerns constructions with rigid bodies. Thus geometry yields propositions about the behaviour of these bodies and is, in this sense, an emperical theory. This standpoint was adopted by H. v. Helmholtz [HEL] and A. Einstein, who wrote:

> "Feste Körper verhalten sich bezüglich ihrer Lagerungsmöglichkeiten wie Körper der euklidischen Geometrie von drei Dimensionen; dann enthalten die Sätze der euklidischen Geometrie Aussagen über das Verhalten praktisch starrer Körper." ([EIN] p. 121)

Consequently, G. Ludwig suggested [LUD2] going one step further and formulating geometry explicitly as a theory of possible operations with practically rigid bodies, using as basic concepts "region", "inclusion" and "transport".

In 1977 I started carrying out this program in detail. One part was completed by connecting the theory of regions and transports with the mathematical results on the Helmholtz-Lie problem [FRE]. Together with a second part dealing with mobility and distance measurement by chains, this approach was presented at a conference in Osnabrück in November 1977 [SCH1]. Following suggestions generated by the discussions at this conference, I added a chapter on operations with rigid bodies, which completed this work. The German version entitled "Zum physikalischen Raumproblem" was presented to the Fachbereich 5, Mathematik/Naturwissenschaften der Universität Osnabrück as the author's Habilitationsschrift and accepted in November 1978.

In conclusion I should like to thank K. Bärwinkel, J. Ehlers, A. Hartkämper, A. Kamlah, G. Ludwig, D. Mayr and G. Süßmann, whose encouragement and interest have been of great value to me. Further I express my gratitude to T. and M. Louton for revising the translation of my manuscript. I have also much appreciated Frau P. Ellrich's and Frau A. Schmidt's rapid and accurate typing of the manuscript.

August 1979

CONTENTS

1. INTRODUCTION

This book presents an axiomatic approach to the foundations of physical geometry. This will be developed with the intention of exploring some problems dealing with physical space.

1.1. Geometry, understood as the theory of physical space (resp. spacetime), plays a constitutive role in physics. Every physical theory contains geometrical concepts (ignoring for the moment some very general versions of thermodynamics or quantum theory). Moreover, the identification of physical concepts such as energy, momentum, angular momentum, occuring in different physical theories can be reduced - via E. Noether's theorem - to the identification of the different concepts of space (resp. spacetime) and the corresponding symmetry groups. If we describe the same nature by different physical theories, there must be a connection between these theories. Geometry is the main medium of such a connection.

Another aspect of the fundamental role of geometry is often formulated as follows : (almost) every physical measurement can be traced back to a geometrical measurement. This notion of "tracing back" can be restated in a more precise manner, using the inter-theoretical relation of a physical theory PT_1, which is a "pre-theory" r.e. another theory, PT_2, under consideration (see [LUD 3]). The experimental "data" in PT_2 consist of theoretical statements of the pre-theory PT_1. These statements in turn are interpretable in terms of basic experiments in PT_1, and so forth. Thus, a systematic construction of physics would consist of a hierarchy of theories and pre-theories where geometry is located at the outset.

Therefore, an axiomatic formulation of geometry as a physical theory is of considerable interest for methodological research, especially for a theory which is a pre-theory for all others but has no pre-theory (at least in the sense indicated above).

Geometry, understood as a physical theory, presents two principal questions:

1. How can the geometrical concepts be interpreted or, in what sense may geometry be applied to "real things"?
2. Where do we know that geometry is "true" (if at all)?

If it is possible to formulate geometry as a physical theory, the second question is reducible to the general question of the validity of a physical theory. Roughly speaking, a theory is accepted as true if it does not conflict with the results of experiments. Of course, this statement is not as trivial as it appears on the surface. For instance, in most cases "conflict" could not be unterstood as "logical contradiction". Moreover, each theory is at most approximately true, and the role of approximation should be investigated in connection with criteria of validity (see the discussion in [LUD 3]).

There are nevertheless other opinions w.r. to the validity of geometry, for instance those which assume geometry being a part of "protophysics", which is thought to be the a-priori base of empirical physics (see [BÖH]). This opinion will be discussed briefly below.

The reference to experiments brings us back to the initial question, whose answer will occupy the remainder of this book. To approach the problem of the empirical content of geometry, 3 scales of dimension need to be distinguished:
The microscopic (μ), the macroscopic (or "laboratory")(L) and the

astronomic (A) dimension. The present axiomatic approach is restricted to the laboratory dimension, from which our geometrical perception arises and in which it works well.

Moreover, we will confine ourselves to that part of L-geometry operating with rigid bodies (e.g., rulers and compasses, or building-stones and joists), since we feel, that this geometry is a pre-theory for example of geometrical optics as well as of other physical geometries.
The geometrical aspects of the microscopic and the astronomic dimensions can only be explored indirectly by means of certain L-experiments (e.g., using micro- or telescopes) and theories. It is doubtful whether the corresponding geometries can be formulated independently of such theories - e.g. of quantum theory or general relativity. In contrast to this we shall formulate the L-geometry without utilizing classical mechanics.

Clearly there is a close connection between the various geometries. The experiments which permit us to deduce the nature of space on either a small or large scale ultimately encounter processes, which take place in the L-dimension. Such processes are described by L-geometry and other "L-theories". Hence L-geometry must be viewed as one of the pre-theories of such theories in whose context μ- and A-geometry occur, and only L-geometry has the characteristics of a universal pretheory as mentioned above.

On the other hand, L-geometry should be possibly viewed as a limit of these more extensive theories, not only due to its mathematical structure, but also due to its rules of interpretation. If we interpret L-geometry as a theory of the assemblage of rigid bodies, this could imply that general relativity together with equations of matter, or, respectively, quantum theory of solid state, could provide the possibility of assembling certain bodies, thus revealing the euclidean structure of space in laboratory dimensions. The solution of such a problem of consistency symbolized by the diagram

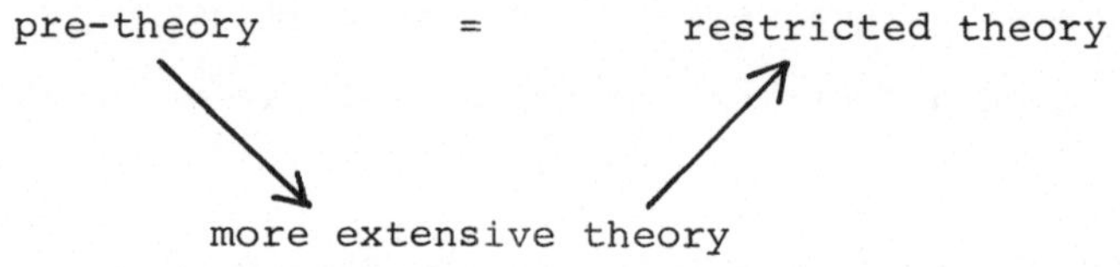

would legitimize and explicate the aforementioned identification of various concepts of space in different theories.

Even in the case of laboratory geometry (from now on just geometry) the relevant concepts - "point", "line", "plane", "distance" and "angle" - have no direct physical meaning. Whereas points may be represented by "small" spots or markings it is more difficult to explain in physical terms what a line or a distance between two points is. Of course, it is possible to consider certain procedures producing straight edges or for comparing distances and to "define" the corresponding concepts operatively by these procedures.

Basically, the proto-physical approach of the Erlangen-Konstanz group proceeds in this way. They formulate standards for measuring devices and so-called "principles of homogeneity", from which they seek to derive a Euclidean geometry (see [BOE] p. 83 ff.).

This approach seems to depreciate the empirical basis of geometry in favour of a normative basis. However, one can argue, that the empirical content of geometry is then manifested in the tacit assumption of practicability of the standards or workability of the procedures.

When one tries to stringently analyze the conditions of geometrical procedures one must translate the primitive geometrical operations into a mathematical language and formulate the conditions of workability as mathematical axioms: the goal of the present volume.

If, for example, one compares distances by transporting measuring rods, these rods must not be deformed during transport. It is not satisfactory to claim: "experience shows that they are not deformed", because deformation would need to be measured by other non-deformed measuring rods.

A universal deformation is not detectable. Hence it is meaningless and does not exist, says the operationalist, distance is what is measured by transporting measuring rods. In principle we agree, but would still try to improve on this argument at two points. First, the conditions of the workability of the proposed operations should be explicitely formulated. One apparent condition in the comparison of distance is, that two measuring rods made from different materials have the same length before some transport, if and only if they have the same length after the transport (see (2318)). When formulated as mathematical axioms, these conditions make it possible to define the concepts under consideration as mathematical terms within the formalized physical theory. This has the additional advantage that we are now no longer restricted to one specific method of measuring a quantity. Each appropriate theorem of the mathematical part of the theory (e.g., on the equivalence of two definitions) now yields another possible "operational definition", which necessarily corresponds to the same physical concept.

This is the second improvement of an operationalism which does not take into account the de facto pluralism of measuring methods. (Admittedly it would be necessary to study other physical theories such as geometrical optics and their connection to rigid body geometry in order to consider the full pluralism of geometrical measurements.)

In short, the above is the standpoint of G. Ludwig (see [LUD 1,2,3]), which we adopt. Given a mathematical formulation of a physical theory, certain sets and relations of the theory play the role of physically interpretable terms; the interpretation being either direct or derived from specific pre-theories. However the development and substantiation of the measuring procedures for the non-interpreted terms is achieved by appropriate mathematical constructions within the theory. If it is possible to derive all terms and theorems of the theory from the interpreted terms by means of some axioms expressible in these terms, one has reached the axiomatic basis of the theory.

Turning back to geometry, we have to decide which terms (sets, relations) of Euclidean geometry of the 3-dimensional space E^3 are suited as interpretable terms, i. e. which terms are as close as can be possible to the physical applications of geometry. Following the approach of G. Ludwig ([LUD 2] II and IV), which in some aspects is due to H. v. Helmholtz (see [HEL]), we choose a class of subsets of E^3, called (spatial) regions, which are experimentally realizable by configurations of fixed bodies. The inclusion of regions would correspond to the realization of "sub-bodies" within these configurations. Finally the group of congruent mappings formed by translations and proper rotations (and their products) could be interpreted as describing the transport of rigid bodies.

The formulation of an axiomatic basis of geometry thus amounts to considering an abstract set R of "regions" endowed with the structure given by a relation $<$ on R ("inclusion of regions") and a set T of relations τ on R ("transport mappings"). Further, the triple $(R,<,T)$ is subject to a series of axioms, which may be thought of as physical laws concerning the fixation and transport of rigid bodies. These axioms have to be choosen in such a way that $(R,<,T)$ can be represented over the space E^3 in the manner indicated above. (Being more precise, this does not hold for $(R,<,T)$, but for some kind of completion of this structure, namely $(\bar{R},\subset,\bar{T})$.)

What does the axiomatic basis tell us about the interpretation of points and distances? We will define points as certain systems of "contracting" regions (= minimal Cauchy prefilters). If $R = \bar{R}$ is the class of bounded open subsets of E^3, the points of E^3 are recovered by this construction.
Equivalently, points could be defined as minimal "limits" of regions (= atoms in $\hat{R}$, see (3313) ff.). In any event a process of idealization is involved and since the correspondence between real things and mathematical entities is at best an approximation (see [LUD 3] § 6), one could (and must) represent "points" by small regions. Although this was previously clear, we now have an instrument to translate each statement concerning points into a statement concerning regions (and transport mappings).

This "operational pluralism" becomes more apparent when we consider the measurement of distances. By dint of the axioms which will be postulated, the space of points P becomes a metric space (up to a factor). However, the construction used to develope the existence theorems does not lead immediately to a method of distance measurement.

One possibility would be to consider sets of points which can be reached by transported regions containing a single fixed point. These sets will form, so to speak, a system of concentric balls, naturally equipped with an inclusion relation and a law of "addition". Proving the axioms of an so-called extensive system allows the set of concentric balls to be mapped onto the non-negative reals. This approach can be found in the work of J. Tits [TIT] and similarly in a recent paper of W. Balzer and A. Kamlah [B&K].

Another method of measuring (proportions of) distances can be found in [LUD 2], IV or in the essay of W. Büchel [BOE].
It considers "chains" between points which consist of overlapping congruent regions, and the minimal number of links of a chain with fixed congruence class which are necessary to connect two points. The quotient of the minimal "lengths" of chains connecting two pairs of points should converge to the quotient of the corresponding distances as the congruent links of the chains become smaller and smaller.
A thorough study of the theory of chains is a main topic of this book. For the moment it is only necessary to point out that different measuring methods are possible the applicability of which is guaranteed by the axioms of the theory.

In addition, to facilitate and establish the interpretation of the basic terms R,<,T we will formulate a pre-theory in which R,<,T are derived from fundamental terms associated with the concepts "process", "inclusion", "body", "reference body". In this pre-theory we will try to give an explicit account of the process of abstraction leading from single bodies fixed at certain reference bodies to spatial regions. Congruence is introduced as a relation between regions which can be represented by the same body.
The whole construction works within some arbitrarily chosen, but

fixed reference system, i. e. a certain equivalence class of reference bodies. It remains a task of a kinematic theory to analyze the relation between different geometries of different reference systems. It is convenient to mention here that the concept of an "inertial system", which is usually thought to be purely kinematic, already exists within the foundations of physical geometry. Roughly speaking, an inertial system may be characterized within the reference systems by the property that the congruence relation of regions does not depend on the representing bodies (see [LUD 2] II § 2). One can observe a material-dependent deformation of bodies in non-inertial systems; therefore one would have to distinguish between say "steel-congruent" and "plastic-congruent". For more details see section 2.1.

1.2. Axiomatization of a physical theory in the sense discussed above requires in most cases a representation theorem of a high degree of complexity, when even provable. The axiomatization of quantum theory, for example, is normally based on the coordinatization theorem in projective geometry. We wish to base the present work on certain mathematical ideas and results which have a long history and concern the axiomatic characterization of a class of geometries.

It is not by chance, that this tradition goes back to a physicist, H. v. Helmholtz, who in 1866 formulated the idea of the intimate connection between the possibility of measuring distances and the free mobility of measuring rods and developed from this idea an axiomatic characterization of the classical geometries [HEL]. Starting with S. Lie, 1890, mathematicians have improved on his ideas and brought them into a rather rigorous form. Since then the problem of characterizing a class of geometries by postulates of mobility on the group of automorphisms, known as the Helmholtz-Lie problem has been treated by a great number of mathematicians. (The bibliography in [FRE] is quite complete.) As one can imagine, the problem has become somewhat independent of its physical origin, a condition which this work will (try to) remedy.

The most general results on the Helmholtz-Lie problem have been obtained by J. Tits [TIT] and by H. Freudenthal [FRE] in 1953 and 1954. They used Lie-algebraic techniques and the theorem of Yamabe in order to drop the assumption of a metric as well as earlier assumptions of differentiability.
The basic concepts now are:
a set of points, a topological structure on it and a group of point transformations. The axioms read in the Freudenthal version:

(121) A complete group of uniform homeomorphisms operates transitively on a connected, locally compact uniform Hausdorff space such that the stability subgroup generates at least one orbit which dissects the space.

This determines a rather small class of geometries and groups (see section 4.3).
The physical relevance of the postulates, however, is not immediately clear. For example, should the completeness of the group or the local compactness of the space be regarded as a matter of purely mathematical convenience or as a law of nature?

Our axiomatic formulation, starting with the (R,<,T)-theory yields a partial answer to this problem. The postulates of (121) are obtained by means of mathematical definitions and constructions, for instance by completion of the group of transport mappings T. So far this is a

question of mathematical convenience. But these constructions can only be performed if certain axioms within the (R,<,T)-theory are assumed. Similarly some of the Freudenthal axioms are expressible as "pre-axioms" in the (R,<,T)-setting. Thus our axiomatization reveals the physical core of the postulates (121). Moreover, this method shows some unexpected connections between properties, which seem completely different at first glance. In this context we draw the readers attention to the double interpretation of axiom R4 (see (317)) as a property of "local pre-compactness" or "pre-transitivity". Another example is theorem (325), which shows the intimate connection between the possibility of measuring distances by means of chains and the Riemannian structure of space.
Hence, our main task consists of finding a pathway which leads from the rather general structure of regions and transport mappings to the Freudenthal system.

1.3. The following is a brief outline of the structure of this work.

In section 2 we develop a pre-theory PT_1 w.r. to the geometry PT_2. As mentioned above, our aim is to give an interpretation of the basic concepts in PT_2, namely "regions", "inclusion", "transport mapping", by means of concepts which are closely related to experimental situations, namely "process", "inclusion (of processes)", "body" and "reference body".

The concept of a "process" could be avoided and really serves only as a pre-requisite for later studies of kinematics. Moreover, section 2 may be considered as an analysis of the axioms R1, R2 in PT_2 (i.e.: (R,<) is a weakly distributive lattice and T a subgroup of Aut R) in terms of the pre-theory PT_1.

In section 3 we formulate the theory PT_2 of regions and transport mappings. The group T is made into a topological group (T,t) in a natural way: a transport is "small", if a certain region meets the transported region. Axiom R3 is the simplest way to insure continuity of multiplication in (T,t). It resembles Freudenthal's axiom of "rigidity". "Points" are now introduced as minimal Cauchy prefilters on $(R,<)$. By axiom R4, a covering law, each region (except the least region O) "contains" at least one point. The set of points P becomes a complete, uniform, locally compact (by R4) Hausdorff space.

Regions may be represented by open, relatively compact subsets of P and transport mappings induce uniform continuous homeomorphisms of P, which are called "congruent mappings". This representation, not in general faithful, generates the lattice $(\tilde{R},\subset)$ and the group $\tilde{T}$. $\tilde{T}$ is completed w.r. to the two-sided uniformity induced by t, thus generating a group $\bar{T}$, also consisting of automorphisms of the uniformity on P. Similarly, $\tilde{R}$ is dense in the lattice $\bar{R}$ of all open, relatively compact subsets of P.

The next steps prepare the proof of Freudenthal's axioms. Unfortunately he postulates completeness of the group w.r. to the uniformity of compact convergence, whereas we had to complete $\tilde{T}$ w.r. to a finer uniformity. Hence we must prove Freudenthal's theorems in this different setting. To this end we postulate axioms R5, a kind of "pre-connectedness" of T, which is mathematically more than we need, but appears physically reasonable.

We define "chains" and compile some simple properties to be used later. Chains of minimal length may serve to define a proportion of distances between points ("chain quotient"), if convergence may be assumed. For this it proves to be irrelevant whether one considers $(R,<,T)$-chains or $(\bar{R},\subset,\bar{T})$-chains. Therefore it is then possible to formulate an equivalent version of axiom R6 on the $(R,<,T)$-level. This is done in section 4. We show that Freudenthal's axiom of mobility is essentially equivalent to the convergence of the chain quotient. In this case P is a Riemannian manifold of a very special type, as is shown in the classification of Tits/Freudenthal. P is essentially isomorphic either to an affine, or hyperbolic, or elliptic space over the real, or complex, or quaternions or Cayley numbers. $\bar{T}$ is isomorphic to the Id-component of the corresponding group of isometries up to some known exceptions.

Whereas this has been known since the fifties, the above mentioned theorem of equivalence provides a new insight into the interdependance between metric and Riemannian structure of physical space and may be viewed as an exact reformulation of the corresponding idea of H. v. Helmholtz.

It is not difficult to construct minimal chains along geodesics in an isotropic Riemannian space. The other direction of the proof is more involved. First, by [YAM] one can find arbitrarily small invariant subgroups N of $\bar{T}$ such that $\bar{T}/N$ is a Lie group and hence P/N a C^{∞}-manifold. The convergence of the chain quotient is carried over in this setting. Further there exists a canonical chart in which the stability subgroup operates by rotations and the transitive transformations operate locally "almost" as translations. Now, if the stability subgroup would not operate transitively on the sphere of constant distance, a chain consisting of long, thin links would be kinked if connected to a point in a "forbidden" direction and the

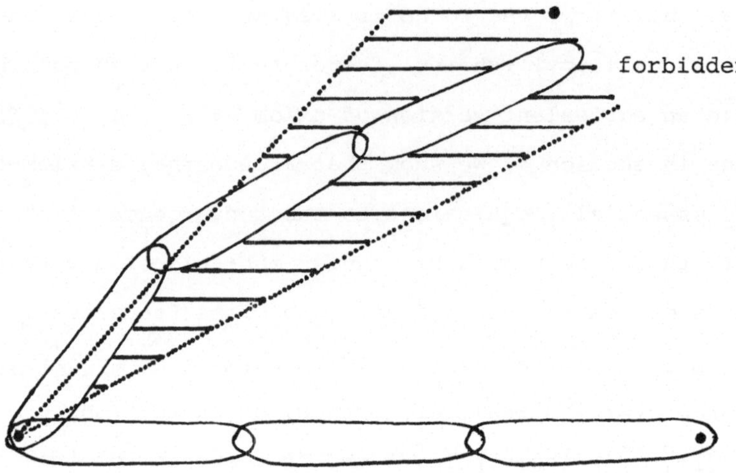

fig. (131)

minimal length of these chains would increase.
Other chains with "globular" links could not be sensitive to a restriction of mobility and thus the chain quotient could depend on the form of the links, contradicting the assumption of convergence.

It is a straightforward matter to separate Euclidean 3-space from the list of geometries obtained by Tits/Freudenthal if one postulates curvature to be 0 and dimension 3. This is done in section 3 by means of axioms R7 and R8 which can be formulated on the (R,<,T)-level. It is known that the dimension of a metric space can be characterized by certain covering properties.

Plastically speaking, each region could be lined with stonework in such a manner that maximally 4 building stones touch each other.

A Riemannian space is known to be flat iff in each triangle the bisecting line of two sides is half as long as the third side.
By the results of section 4 this is expressible as a property of minimal chains.

1.4. How can these results be applied and what sort of new problems arise from them?

First the achievement of an axiomatic physical geometry furnishes an example to test some methodological definitions proposed in [LUD 3] § 10.5. It concerns, among others, the question, whether points and distances for example may be viewed as "physically realizable facts" (reale Sachverhalte) in the sense defined therein, and which limit processes need to be considered for this purpose. Equally interesting, the problem of the appropriate uniformity of physical uncertainty has not been tackled.

Further, it would be desirable to extend the present approach to space-time-geometry. Here it is possible to proceed in various directions. One way would be to consider a lattice of spatio-temporal regions (i.e. "processes") and a group of "reproductions" as basic notions and to use the results on the generalized Helmholtz-Lie problem leading to pseudo-Riemannian geometries [FRE 2]. This has recently been achieved by D. Mayr [MAY]. Another possibility is to obtain space-time-geometry by pasting together the geometries of different inertial frames by means of coincidences. The author is currently working on this aspect [SCH 2].
If we had insights as to, why spacetime is a smooth manifold, perhaps we could better understand the physical reason of the assumption of differentiability in classical mechanics (see [LUD 2] p. 392).

Besides the geometrical measurements of rigid body (RB) L-geometry, experimentalists use a wide variety of other methods from electron microscopy to radio astronomy in order to explore geometrical extensions of objects. From a methodological point of view this is a question of indirect measurements of geometrical entities by means of physical laws of the corresponding theory and an interpretation of certain basic concepts by a pre-theory (here L-geometry is introduced). The simplest theory in which these relations could be analyzed in detail would probably be geometrical optics. The pattern of intensity in the image plane of a microscope can be interpreted in terms of RBL-geometry. By the laws of lens refraction we may infer some geometrical structures in the object plane which are far below the scale of accuracy of RBL-geometry (otherwise the microscope would be useless). For instance the validity of Euclidean geometry on a small scale could thus be partially understood by means of an axiomatic formulation of geometrical optics. More difficult would be the analogous analysis in case of quantum theoretical, indirect measurements of geometrical entities. Usually (non-relativistic) quantum theory is coupled to geometry by some unitary representation (up to a factor) of the Galilei group, which yields among others the position observable. The physical significance of such a representation is well established by considering transports and movements of preparation and measurement apparatus. But this does not define the geometry on a microscopic scale because the apparatus cannot be adjusted within an arbitrary accuracy. If a length of for instance 10^{-15} m is the result of an experiment, it is calculated by means of quantum theory (and other theories), e. g. the total cross section in a scattering experiment.

The reconstruction of space(time) structure on a microscopic scale in terms of scattering experiments should be facilated by the fact that an axiomatic analysis of the general form of quantum theory as a theory of macroscopic devices (= preparation and measurement apparatus) has already been achieved (see [LUD 1]. Nearly everybody believes, that the space(time) on a very small scale is different from the Euclidean one, but we do not even know to what extent it is really Euclidean on atomic scales.

2. OPERATIONS WITH RIGID BODIES

2.1. GENERAL EXPLICATION

The primitive concepts of the theory developed in section 3 - region, inclusion, transport mapping - are very intuitive and, at first glance, seem to be appropriate for direct physical interpretation: Spatial regions are occupied by solid, rigid bodies "at rest", the inclusion of regions corresponds to the inclusion of "sub-bodies" and transport mappings are thought to represent the transition between two regions occupied successively by the same body (or its copies).

A closer inspection however reveals that these "primitive" concepts are theoretical constructs obtained from the level of rigid bodies via an intricate process of abstraction. It may even be suspected that this process presupposes the use of geometry, in which case the attempt to found a physical geometry would become a vicious circle. Therefore it will be necessary to give an explicite, mathematical formulation of this process. The construction of regions and transport mappings depends on certain conditions, which appear as the laws of operations with rigid bodies. These are the axioms $\pi 1$, $\pi 2$, B1 to B8 and J1.
In this section we will explain the meaning of the concepts which are now considered as primitive and, further, illustrate the ideas leading to the formal definitions and axioms of part 2.2 and 2.3.

"Rigid" bodies are for example: stones, walls, nails, boards, bricks. Under weak forces they do not exhibit perceptible deformation or abrasion. Thus a droplet of water or honey, a cloud of steam, a rope, a chain or a pile of straw are not rigid bodies. If the forces exerted on a body are carefully restricted in order to avoid deformations, this body also may be regarded as rigid, for example a pair of compasses or a table together with a steel ball lying on it. This concept of rigidity is a pre-theoretical concept of everyday experience. We are not concerned with the idealization of the rigid body concept in classical mechanics or with the (non-existing) rigid body in relativity.

Neither are we looking for an absolute definition of the rigid body which allows for all effects as elastic deformations by small forces, heat expansion, magnetostriction etc. (see for instance [REI] § 18). This is regarded as an issue of refining rigid-body-geometry within the framework of richer theories. The first step in constituting the geometry does not analyze these effects; but of course they restrict the degree of accuracy of this kind of geometry. Finally, it would be possible to exclude all rigid bodies which are produced using geometry, say with straight edges or spherical shapes. These are not needed for the foundations of geometry. After the complete theory has been set up, it is possible to construct such symmetric objects, but they do not play the central rôle as in the approach of "protophysics" [BOE].

Next we turn to the notion of bodies "at rest". Obviously "rest" can only be defined with respect to other bodies and there exist different kinds of rest according to different systems of reference. Two bodies are in relative rest in a situation where they can be considered as parts of a rigid body. But this relation fails to be transitive in reality, thus "rest" cannot be defined only in terms of "rigid bodies" and "inclusion". If the transporting operations we consider are to be performed, almost every body must be moved. Hence, there must be a criterion for the body being at rest after the transport. To this end we suggest to distinguish a subset of bodies which are not be moved during the experiments to serve as references for other bodies in order to define the state of rest. We will call them "reference bodies" ("Gerüste" in [LUD2] p. 26 ff.) and say that a body is at rest, if it is fixed with respect to some reference body, i.e., if it is part of a larger rigid body (a "configuration") which also contains the reference body. Consider for example the walls of a laboratory or of a space-ship cabin. The mass of these reference bodies must be so large that the small forces, which occur when fixing or loosening the configurations, practically do not affect their movement.

For the mathematical description of our basic notions we will consider a set B (for body) endowed with the structure given by a relation $\sqsubset$ (for inclusion) on B and a subset $R \subset B$ (for reference bodies) fulfilling several axioms. To make it clear, B is not the "set of all rigid bodies" which is a rather ill-defined object. B is a term within a mathematical theory in the sense of [BTS]. An assemblage such as $k \in B$, which may be written down in this theory, is additionally conceived as a translation of the sentence "this object, which I will call k, is a rigid body" (according to the non-formalized criteria mentioned before). Similarly, if one screws two boards b_1, b_2 together, forming a body b, one may write "$b_1 \sqsubset b$ and $b_2 \sqsubset b$". Usually a body b has been artificially composed of its minimal sub-bodies, so that it is entirely clear which bodies satisfy "$k \sqsubset b$" and "not $k \sqsubset b$".

"Not $k \sqsubset b$" means that the body b is constructed without using k as a part of it; it does not mean that the body b which contains k is actually decomposed. (This is, by the way, granted by the reduction of $\sqsubset$ to the inclusion of processes given below.) Moreover, the relation $\sqsubset$ does not apply, if for example a piece of wood is cut into two parts. Sticking together the parts constitutes a new body different from the original piece of wood.

The crucial point in this approach is the possibility of taking compound bodies to pieces and to reassemble. The geometer has to decide whether he has reproduced the same body with which he started (without using geometry), analogous to the mathematician who must recognize a sequence of symbols a second time.

We will introduce another fundamental concept, the concept of a "process" $p \in \pi$, which is not, strictly speaking, necessary in the present context, but which is convenient for connecting this geometry with other physical theories. The fixing of a body in some configuration, the movement of a body between two distinguished stages, e.g. between the points of return of a pendulum, are examples of "processes". We will use this notion in the sense of an individual event, happening but once, which would be represented by a space-time region. Again, there is a natural concept of spatio-temporal inclusion $\sqsubset$. A body now corresponds to a subset of processes, namely the stages of its history, or - using a space-time diagram - the sections of its world-tube. Note that the sub-processes corresponding to sub-bodies do not belong to this subset $k \subset \pi$ (see (223)).
Formally, the inclusion of bodies $\sqsubset$ can be defined in terms of $\sqsubset$. Thus our final formulation is as follows: the set π is endowed with the structure $(\sqsubset, B, R)$. In principle, physical geometry, as presented in this book, could be formulated in these terms. (In spite of that we would prefer to use some kind of abbreviation.) Moreover, the author is convinced that any physical theory could be represented as an - admittedly very complicated - structure on π.

We will now discuss the problem how spatial regions can be defined in terms of configurations $l \in B_k$ containing the body $k \in B$. First note that the definition will depend on the chosen "frame" of reference which may be defined as an equivalence class of reference bodies: two reference bodies within the same frame $\varphi \in \mathcal{F}$ could, in principle, be connected rigidly. The "position" of a body k within a configuration $l \in B_k$ does not change if some other body is removed from l or added to l. Considering a finite number of constructive alterations we obtain "chains" of configurations of k, say $l_1 \sqsubset l_2 \sqsupset l_3 \sqsubset \ldots \sqsupset l_N$, and in this case all l_i are said to belong to the same "position" of k.

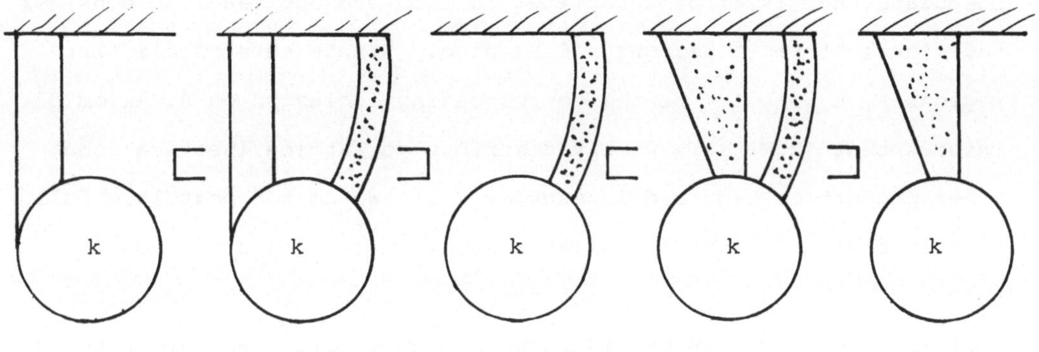

(The hatched region represents the reference body)

fig. (211)

Now, the crucial problem is to decide whether two positions of different bodies occupy the same region $A \in \mathcal{R}$. We have to abstract from the material properties of the bodies - as colour, mass, hardness - exept the property of geometrical extension. But this property should be expressible in our approach in terms of composition of configurations. Two positions are equivalent with respect to the considered constructions if they can be substituted one for another in each configuration. A "substitution" in this context can be defined by means of the basic notions already provided. The proof that we in fact obtain an equivalence relation in this way raises some difficulties and we have to formulate some simple and intuitively clear assumptions as axioms (B1 to B4). The inclusion $\sqsubset$ of bodies in a natural way induces an inclusion $\prec$ of regions. Whereas a body contains only a finite number of sub-bodies, a region may be infinitely divisible (in an appropriate mathematical model), since it may be represented by a sequence of positions containing smaller and smaller sub-bodies.

Regions must be considered as connected (although this will not be proved in a strict sense) and therefore it would not be realistic to postulate that $(R,\prec)$ be a lattice. To this end one needs to consider additional "formal" regions $a \in R$, namely finite sets of disjoint regions $A_i \in R$. Let $<$ be the corresponding inclusion on R. Axiom B5 insures that $(R,<)$ is a weakly distributive lattice (besides some other properties required in section 2.3). We do not postulate full distributivity for the following reason: if regions are to be (in section 3) represented by point sets, two sets with the same topological closure cannot be distinguished physically, but distributivity depends on a certain choice of representation. (The counter-example (329) illustrates this point.)

The construction of transport mappings in section 2.3 does not require further basic concepts. Two regions are said to be "congruent" if they can be represented by the same body. At first glance it seems that a pair of congruent regions (A_1,A_2) should determine a transport mapping $\tau : R \to R$ such that $\tau A_1 = A_2$. But this assignement is not unique since $\tau A = A$ is satisfied also by a symmetry of A, not only by the identity. Therefore the transport mapping will depend explicitely on the representing positions of the body k: $\tau = T(\pi_1,\pi_2,k)$. It is implicitely assumed that each body exhibits small inhomogenuities, for instance in colour, such that any non-trivial symmetry transport yields a new position, even if its region is unchanged.

Now we will define $T(\pi_1,\pi_2,k)\ B_1$, $B_1 \in R$, operationally. Imagine for instance an iron block k, fixed on a table in some position π_1. Now the block is loosened and rotated around some axis perpendicular to the table by an angle of, say, 30°. Then it is again fixed and its new position will be π_2.

For simplification we will first assume, that the region B_1 is disjoint from π_1 and thus may be represented by another block 1, say lying on the table "simultaneously" with k.

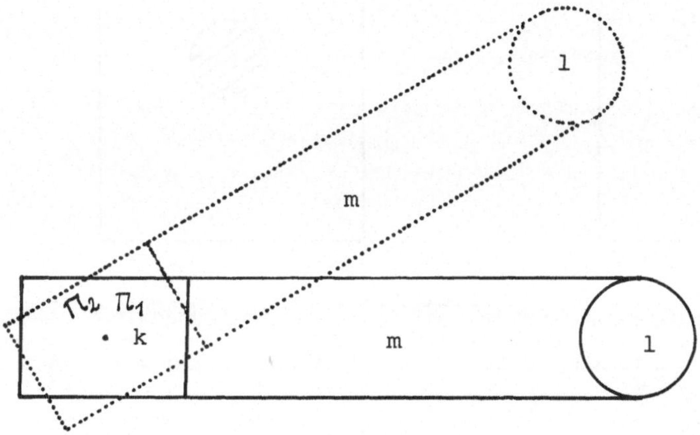

fig. (212)

Now we connect k and l rigidly, say by an iron band and clamp frames, thus constructing a new body m $\sqsupset$ k, l and then we move m in such a manner that subsequently k will be in position π_2. Then l in its new position will represent the region $B_2 \underset{def}{=} T(\pi_1,\pi_2,k)\ B_1$.

At this point several questions arise:

1. Obviously we have used an assumption of the following form: if a body k can be moved between two positions π_1 and π_2, it can be moved in the same way when considered as part of a larger body m. Enlargement of a body does not restrict its mobility. We consider this assumption not simply as an axiom, but rather as a criterion for "inertial frames" (see [LUD2] p. 28). A material-dependent deformation occuring in non-inertial frames would hinder the required mobility, as exemplified in the figure below.

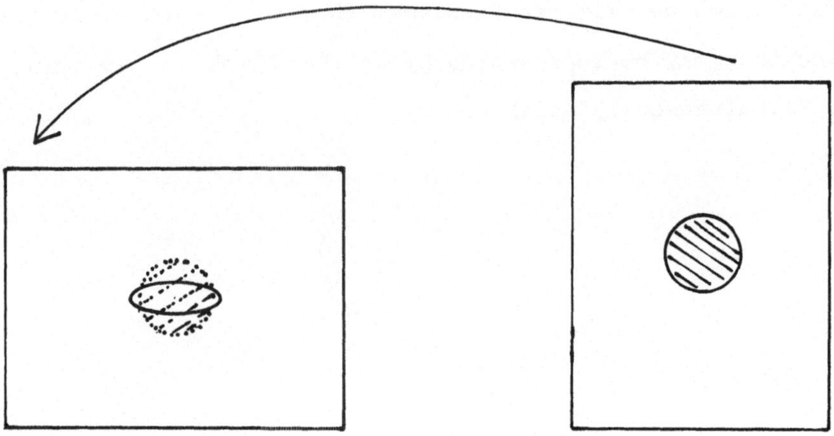

fig. (213)

Of course, in an approximately inertial system one could keep this effect small by using very rigid materials. This shows that the 3 entities rigidity, inertiality and accuracy are intertwined. But nevertheless it is possible to test inertiality by means of purely geometrical experiments and non-inertiality forms a strict limit to geometry. We have to postulate that there exist inertial frames $\varphi \in J$ and the further development of geometry is restricted to such frames.

2. We have to show that $\tau = T(\pi_1, \pi_2, k) : R \to R$ is well-defined. First, the region $B_2 = \tau B_1$ must not depend on the rigid connection forming m and on the position representing B_1. In general, B_1 will not be disjoint from the region of π_1 and in this case we have to consider an auxiliary region C_1 disjoint from both and perform the above construction twice. Again the result must be independent of C_1. The various proofs are based on 3 additional, simple axioms B6, B7, B8 concerning configurations.

The remaining task is to show that the class of transport mappings τ obtained in this way forms a group of $(R,\prec)$-automorphisms. Then the same holds for the canonical extensions on $(R,<)$.

We will finish this section with some remarks on the object and interpretations of these constructions. Of course, there is no need for axiomatization of this elementary field of experience caused by difficulties of applying geometry to the construction of buildings, furniture or physical apparatus. The only point is to show how the foundations of geometry could in principle be accomplished. Therefore the axiom system presented is not analyzed in view of consistency, completeness and independence. Surely not all propositions which are intuitively true for operations with rigid bodies can be proved from our axioms. Only those propositions which are used to derive the basic concepts and axioms R1, R2 of section 3 are considered and compressed into the 11 stated axioms. Obviously, we had in mind Euclidean geometry, and at least one axiom, (2234) (iv), is not suited for compact spaces (spherical, elliptic geometry).

However, a model which satisfies our axioms and thus shows their consistency would be desirable for another reason. We have (directly or indirectly) postulated a great variety of regions. That means that there must exist a great number of configurations, hence of spatio-temporal processes and it is not clear a priori whether ordinary space-time is large enough to contain the required variety of processes.

2.2 CONSTRUCTION OF REGIONS

We now turn to the mathematical details of the "basic geometry", PT_1. Its species of structure is given by

(221) principal base set: π, structural term:

$(\boldsymbol{\sqsubset}, B, R) \in P(\pi \times \pi) \times PP\pi \times PP\pi$,

and an axiomatic relation: $\pi 1, \pi 2$, B1 to B8, J1.

(The terminology agrees with [BTS] or [LUD 1] p. 81 ff.)

The elements $P \in \pi$, are called <u>processes</u>; $P \boldsymbol{\sqsubset} Q$ is read as "P is a part of Q". The elements of B are called <u>bodies</u>. A body $k \in B$ is identified with the set of its movements, hence with a set of processes: $k \subset \pi$. $R \subset B$ is the subset of <u>reference</u> <u>bodies</u> ("Gerüste").

(222) <u>Axiom $\pi 1$:</u>

(i) $\boldsymbol{\sqsubset}$ is a partial ordering on π,

(ii) B is a family of disjoint non-empty subsets of π,

(iii) R is a non-empty subset of B.

Two bodies k,l can be rigidly affixed in order to form a new body m. Thus k (and l) will be called a part of m: $k, l \sqsubset m$. In order to clarify the relation between $\sqsubset$ and $\boldsymbol{\sqsubset}$ and to motivate the exact definition of $\sqsubset$, the following example has been provided:

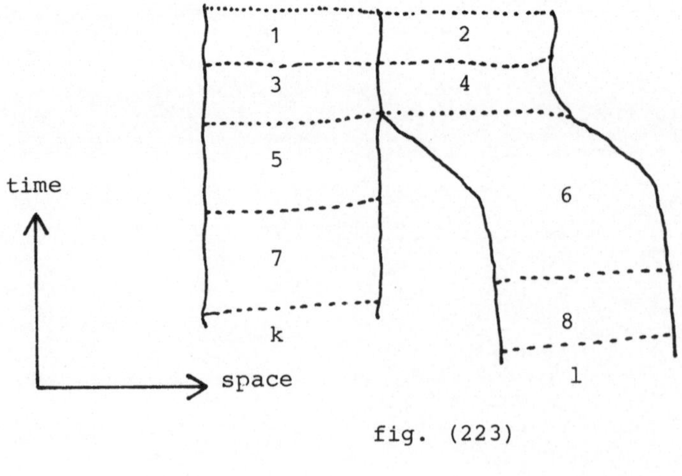

fig. (223)

The "minimal" processes 1,...8 and their "unions" 13, 135, etc... are represented as subsets of 2-dimensional space-time. The 3 bodies are identified with the subsets:

k = {1,3,5,7,13,35,57,135,357,1357},

l = {2,4,6,8,24,46,68,246,468,2468},

m = {12,34,1234}.

The body k is a part of m in the sense that each process $P \in m$ contains a subprocess $Q \in k$, namely $12 \sqsupset 1$, $34 \sqsupset 3$, $1234 \sqsupset 13$. This leads to the following

(224) Definition: Let be k, m ∈ B.

$$k \sqsubset m \underset{\text{def}}{\iff} \forall P \in m \; \exists Q \in k \text{ such that } Q \sqsubset P.$$

$\sqsubset$ can be proved to be a partial ordering on B, provided that the following axiom holds.

(225) <u>Axiom π2:</u> Let be $P \in \pi$, then the set

$$\{Q \in \pi \mid Q \sqsubset P\} \text{ is finite.}$$

(226) Proposition: $\sqsubset$ is a partial ordering on B.

Proof: Clearly, $\sqsubset$ is reflexive and transitive. In order to prove the antisymmetry assume $k \sqsubset m$ and $m \sqsubset k$ and $P_o \in k$. There exists some $Q_o \in m$ such that $Q_o \sqsubset P_o$, further some $P_1 \in k$ such that $P_1 \sqsubset Q_o$, etc... . We inductively construct a countable infinite sequence $P_o \sqsupset Q_o \sqsupset P_1 \sqsupset Q_1 \sqsupset P_2 \sqsupset \ldots$, which, by axiom π2, will be constant after finitely many steps. Hence $P_n = Q_n$ for some $n \in \mathbb{N}$, and since k and m are either disjoint or identical (by axiom π1 (ii)), we conclude: k = m. □

(227) Definition: Let k ∈ B.

M(k) will denote the set of <u>minimal</u> sub-bodies of k, that is:

$$M(k) \underset{\text{def}}{=} \{\, l \in B \mid l \sqsubset k \text{ and } \forall m \in B,\ m \sqsubset l \Rightarrow m = l\}$$

The following proposition is immediate.

(228) Proposition: For each $k \in B$,

$\{l \in B \mid l \sqsubset k\}$ is finite,

$M(k)$ is finite and non-empty.

(229) Definition: Let be $k, l \in B$.

$k \,\Delta\, l \underset{\text{def}}{\iff}$ there exists no $n \in B$ such that $n \sqsubset k$ and $n \sqsubset l$.

In this case k and l are called <u>$\sqsubset$-disjoint</u> or simply: disjoint.

(2210) Definition:

(i) $B_R \underset{\text{def}}{=} \{l \in B \mid \exists r \in R \text{ such that } r \sqsubset l\}$

is called the set of <u>configurations</u>.

(ii) Let $\overline{\overline{\Pi}}$ be the equivalence relation generated by the relation $\sqsubset$ on B_R. The set of equivalence classes, $F \underset{\text{def}}{=} B_R/\overline{\overline{\Pi}}$, is called the set of <u>frames</u>.

(iii) Let be $k \in B$. $B_k \underset{\text{def}}{=} \{l \in B_R \mid k \sqsubset l \text{ and } \forall r \in R,$ $r \sqsubset l \Rightarrow k \,\Delta\, r\}$ is called the set of <u>configurations of k</u>.

(iv) Let be $k \in B, \varphi \in F$, and $\overline{\Pi}_{k,\varphi}$ the equivalence relation generated by the relation $\sqsubset$ on $B_k \cap \varphi$. The set of equivalence classes $Pos_\varphi(k) \underset{\text{def}}{=} B_k \cap \varphi / \overline{\Pi}_{k,\varphi}$ is called the set of <u>positions of k in the frame φ</u>. The canonical surjection

$B_k \cap \varphi \to B_k \cap \varphi / \overline{\Pi}_{k,\varphi}$

will be denoted by $l \to pos(l,k)$, the "position of k within the configuration l". (The frame φ is uniquely determined by l and need not be indicated.)

(v) $Pos_\varphi \underset{\text{def}}{=} \{(\pi,k) \mid k \in B, \pi \in Pos_\varphi(k)\}$

is called the set of <u>positions</u> (in the frame φ).

Further, $\overline{pos}(l,k) \underset{\text{def}}{=} (pos(l,k),k)$.

(vi) Let $(\pi_i, k_i) \in Pos_\varphi$ for $i = 1,2$.

$(\pi_1, k_1) \sqsubset (\pi_2, k_2) \underset{def}{\Longleftrightarrow} k_1 \sqsubset k_2$ and $\pi_2 \subset \pi_1$

(or, equivalently, $\pi_1 \cap \pi_2 \neq \emptyset$).

Clearly, $\sqsubset$ is a partial ordering on Pos_φ.

Thus a configuration is thought to be some construction based on a reference body. Two configurations which can be rigidly tied, are relatively at rest and lie in the same frame. Of course, a configuration containing a sub-body k is a configuration of k. (For technical reasons k must be $\sqsubset$-disjoint from each reference body.) If we make constructive alterations in a configuration of k, the position of k remains constant.

In sequel the frame $\varphi \in \mathcal{F}$ is kept fixed and will not be explicitly mentioned in most cases.

Two bodies with the same set of minimal sub-bodies need not be identical since the sub-bodies can be combined in various manners. If, however, the sub-bodies are fixed in definite positions, this ambiguity should disappear. To this end we postulate:

(2211) <u>Axiom B1:</u> Let be $l_i \in B_{m_i}$, $i = 1,2$, and

$M(m_1) = \{n_1, \ldots n_L\}$, further:

$\forall j = 1 \ldots L$, $n_j \sqsubset m_2$ and $pos(l_1, n_j) = pos(l_2, n_j)$.

Then $m_1 \sqsubset m_2$ holds.

(2212) Proposition: If, in addition to the assumptions of (2211), $M(m_1) = M(m_2)$ holds, then $m_1 = m_2$.

Proof: By (2211) we conclude $m_1 \sqsubset m_2$ and $m_2 \sqsubset m_1$. □

If two positions of a body coincide partially, they should coincide totally:

(2213) <u>Axiom B2:</u> Let be $l_i \in B_k$, $i = 1,2$, $n \sqsubset k$ and $pos(l_1, n) = pos(l_2, n)$. Then $pos(l_1, k) = pos(l_2, k)$ holds.

We now turn to the definition of "spatial regions". As indicated in section 2.1, two bodies (in their respective positions) will determine the same spatial region, if, for each configuration belonging to their position, one can be substitituted by the other.

(2214) Definition:

(i) Let $l \in B_k$, $C(l) \underset{\text{def}}{=} \{m \in B \mid l \in B_m\}$.

Hence $C(l) = \{m \in B \mid m \sqsubset l \text{ and } \forall r \in R,\ r \sqsubset l \Rightarrow r \,\Delta\, m\}$.

$C(l,k) \underset{\text{def}}{=} \{m \in C(l) \mid m \sqsupset k \text{ or } m \,\Delta\, k\}$.

(ii) A <u>substitution</u> is defined to be a 6-tuple

$(l_1, k_1, \pi_2, \alpha, l_2, k_2)$

satisfying:

a) $l_i \in B_{k_i}$ for $i = 1,2$,

b) $\pi_2 = \text{pos}(l_2, k_2)$,

c) $\alpha\colon C(l_1,k_1) \to C(l_2,k_2)$ is a surjective mapping,

d) $\alpha\, k_1 = k_2$,

e) $\forall m \in C(l_1,k_1),\ m \,\Delta\, k_1 \Rightarrow$

$(\alpha m = m \text{ and } \text{pos}(l_1,m) = \text{pos}(l_2,m)$,

f) $\alpha m \sqsubset \alpha n \Leftrightarrow m \sqsubset n$.

By f), α is injective.

In order to illustrate the above definition, we provide the following

(2215) Example:

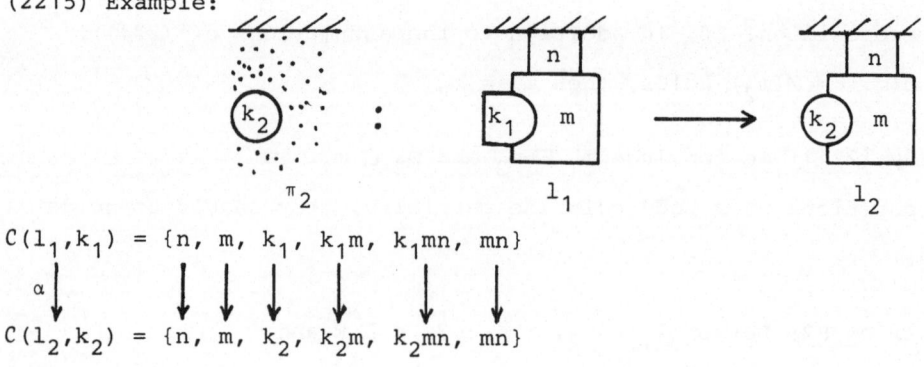

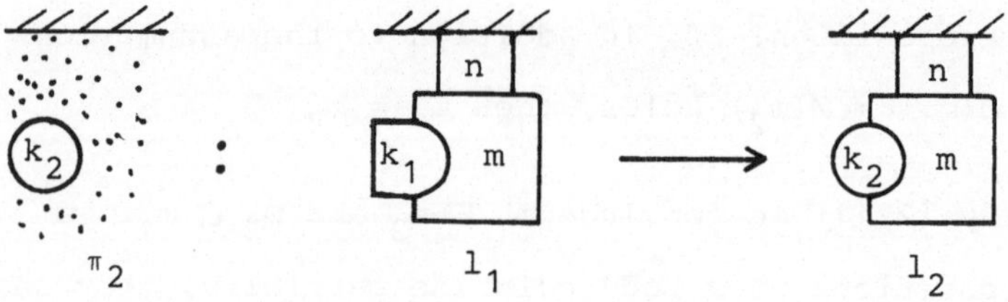

$$
\begin{array}{l}
C(l_1,k_1) = \{n,\ m,\ k_1,\ k_1m,\ k_1mn,\ mn\} \\
\quad \downarrow \alpha \qquad\quad \downarrow \quad \downarrow \quad \downarrow \quad \downarrow \quad \downarrow \quad \downarrow \\
C(l_2,k_2) = \{n,\ m,\ k_2,\ k_2m,\ k_2mn,\ mn\}
\end{array}
$$

(2216) Lemma: Let be $l_1 \in B_{m_1}$, $k_1 \sqsubset m_1$ and $(l_1,k_1,\pi_2,\alpha,l_2,k_2)$ a substitution. Then $C(l_1,m_1) \subset C(l_1,k_1)$ holds. If $m_2 \underset{def}{=} \alpha m_1$, $\tilde{\alpha} \underset{def}{=} \alpha | C(l_1,m_1)$ and $\mu_2 \underset{def}{=} pos(l_2,m_2)$, then $(l_1,m_1,\mu_2,\tilde{\alpha},l_2,m_2)$ is a substitution.

Proof: Let be $m \in C(l_1,m_1)$. $m_1 \sqsubset m$ implies $k_1 \sqsubset m$ and $m \,\Delta\, m_1$, $m \,\Delta\, r \in R$ implies $m \,\Delta\, k_1$, r. Hence $m \in C(l_1,k_1)$. $\tilde{\alpha}$ is surjective, since for all $m \in C(l_1)$: $\alpha m \sqsupset m_2 = \alpha m_1 \Rightarrow m \sqsupset m_1 \Rightarrow m \in C(l_1,m_1)$ and $\alpha m \,\Delta\, m_2 = \alpha m_1 \Rightarrow m \,\Delta\, m_1 \Rightarrow m \in C(l_1,m_1)$.

The other defining properties of a substitution are immediately clear. □

(2217) Definition: Let $S = (l_1,k_1,\pi_2,\alpha,l_2,k_2)$ and $T = (l_2,k_2,\pi_3,\beta,l_3,k_3)$ be substitutions and $\pi_1 \underset{def}{=} pos(l_1,k_1)$.

(i) $T \circ S \underset{def}{=} (l_1,k_1,\pi_3,\beta\alpha,l_3k_3)$, where $\beta\alpha$ is the product of β and α, considered as relations.

(ii) $S^{-1} \underset{def}{=} (l_2,k_2,\pi_1,\alpha^{-1},l_1,k_1)$, where α^{-1} is the inverse of the relation α.

(iii) $E(l_1,k_1) \underset{def}{=} (l_1,k_1,\pi_1,\mathbb{1},l_1,k_1)$, where $\mathbb{1}$ is the identity on $C(l_1,k_1)$.

(2218) Proposition: The class of substitutions has the structure of a groupoid w.r. to the multiplication defined above. The inverse and neutral elements are given by (2217) (ii) and (iii).

We omit the proof, which is straight forward.

(2219) Definition: Let $\pi_i \in Pos(k_i)$ for i = 1,2.

(i) $\pi_1 \sim \pi_2 \underset{def}{\Longleftrightarrow} \forall l_1 \in \pi_1 \; \exists l_2 \in \pi_2 \; \exists \alpha$

such that

$\{\forall n \in B, (n \sqsubset l_1 \text{ and } n \Delta k_1) \Rightarrow n \not\sqsubset k_2\} \Rightarrow$

$(l_1,k_1,\pi_2,\alpha,l_2,k_2)$ is a substitution.

We will say: π_2 can be <u>substituted</u> for π_1. {...} will be called the <u>substitution condition</u> (s.c.)

(ii) $\pi_1 \sim \pi_2 \iff \pi_1 \rightsquigarrow \pi_2$ and $\pi_2 \rightsquigarrow \pi_1$.

π_1 and π_2 will be called equivalent positions (anticipating the result of (2227)). If $\alpha_i = (\pi_1,k_i) \in \mathrm{Pos}$ for $i = 1,2$, and $\pi_1 \sim \pi_2$ we will also write $\alpha_1 \sim \alpha_2$.

(2220) Lemma: It is always sufficient to check the s.c. for minimal sub-bodies n of l_1.

Proof: Let $m \in B$ be sucht that $m \sqsubset l_1$ and $m \,\Delta\, k_1$. Then there exists a minimal $n \in B$, $n \sqsubset m$, enjoying the same properties. Hence $n \not\sqsubset k_2$ and, consequently, $m \not\sqsubset k_2$. □

It is reasonable to postulate mutual substitutability for equivalent positions only under the condition, that no parts of k_2 are used to fixate k_1 and vice versa, hence the s.c. in (2219)(i). This trivial complication has to be considered in some of the following proofs. Therefore we want to replace the bothersome sub-bodies with "copies" (i.e. bodies with equivalent positions) and postulate:

(2221) <u>Axiom B3:</u> Let $(\pi,k) \in \mathrm{Pos}_\varphi$ and $L \subset B$ be a finite subset. Then there exists a position $(\pi',k') \in \mathrm{Pos}_\varphi$ such that $\pi \sim \pi'$ and $(\forall n \in B,\ n \sqsubset k' \Rightarrow n \notin L)$.

The following lemmata serve to prove (2227), namely that $\sim$ is an equivalence relation when one takes into account the complication mentioned above. These are technical in nature and thus may be omitted, without loss of continuity, by the reader who is not interested in these details.

(2222) Lemma: Let $l \in B_{k_2}$, $k_1 \sqsubset k_2$, $\kappa_i = pos(l,k_1)$ for $i = 1,2$, $\mu_1 \in Pos(m_1)$, $\kappa_1 \sim \mu_1$, $(l,k_1,\mu_1,\alpha,l',m_1)$ be a substitution, $m_2 = \alpha\, k_2$ and $\mu_2 = pos(l',m_2)$. Then $\kappa_2 \sim \mu_2$ holds.

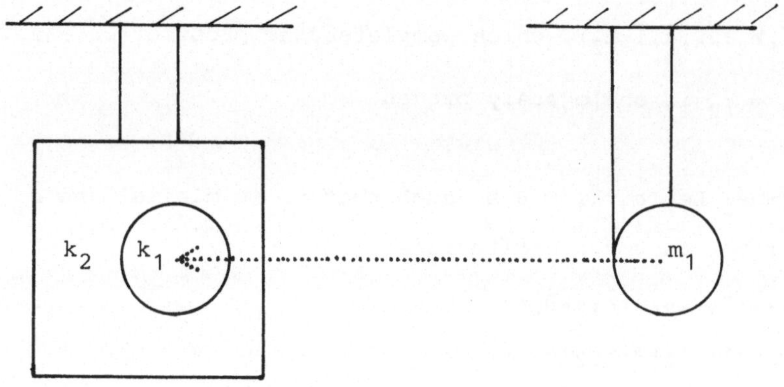

fig. (2223)

Proof:

1. "$\kappa_2 \sim \mu_2$".

Let $\tilde{l}$ be a configuration $\tilde{l} \in \kappa_2$ and the s.c. be satisfied: $\forall p \in B$, $(p \sqsubset \tilde{l}$ and $p \Delta k_2) \Rightarrow p \not\sqsubset m_2$.

Note, that $k_1 \sqsubset k_2$ implies $m_1 = \alpha\, k_1 \sqsubset \alpha\, k_2 = m_2$.

1.1. We assert: $\forall p \in B$ such that p is minimal, $p \sqsubset \tilde{l}$ and $p \Delta k_1$ follows: $p \not\sqsubset m_1$.

Proof:

1.1.1. Consider the case $p \Delta k_2$. Now $p \not\sqsubset m_2$ by the above s.c., hence $p \not\sqsubset m_1$.

1.1.2. In the case $p \sqsubset k_2$ we infer from $p \Delta k_1$ that $p = \alpha p \Delta \alpha\, k_1 = m_1$, hence $p \not\sqsubset m_1$.

1.2. Since $\kappa_1 \sim \mu_1$ and the s.c. is proved for minimal p (see (2220)), there exists a substitution $(\tilde{l},k_1,\mu_1,\beta,\bar{l},m_1)$.

1.3. We assert: $\beta k_2 = \alpha k_2 (= m_2)$.

Proof: We write $M(k_2) = M(k_1) \,\dot\cup\, R$ and conclude

$M(\alpha k_2) = M(m_1) \dot{\cup} R = M(\beta k_2)$ since $\forall m \in R$, $\alpha m = \beta m = m$. Further, $pos(\bar{l},m) = pos(l',m)$ and $pos(\bar{l},m_1) = pos(l',m_1)$ because $\bar{l},l' \in \mu_1$. The latter implies $\forall\ n \in M(m_1)$, $pos(\bar{l},n) = pos(l',m)$ by definition (2210) (iv). By (2212), $\alpha k_2 = \beta k_2$ follows.

1.2. By virtue of (2216) there exists a substitution $(\tilde{l},k_2,\mu_2,\bar{\beta},\bar{l},m_2)$, which completes the proof of $\kappa_2 \sim \mu_2$.

2. $\mu_2 \sim \kappa_2$ is analogously proved. □

(2223) Lemma: Let $n_1,k_1 \in B$ be such that n_1 is minimal and $n_1\ \Delta\ k_1$, further assume substitutions

$(l_{11},n_1,\nu_2,\alpha,l_{12},n_2)$,

$(l_{12},k_1,\kappa_2,\beta,l_{22},k_2)$,

$(l_{22},n_2,\nu_1,\gamma,l_{21},n_1)$.

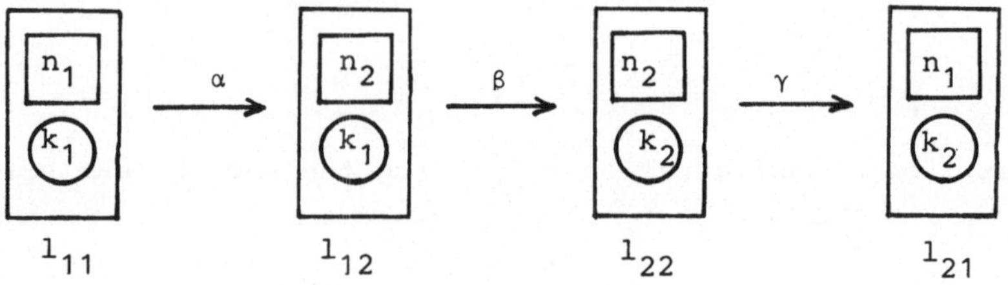

fig. (2224)

Then the following holds:

(i) $C(l_{11},k_1) \subset C(l_{11},n_1)$ and

$C(l_{21},k_2) \subset C(l_{21},n_1)$.

(ii) Set $D_1 \underset{def}{=} C(l_{12},k_1) \cap C(l_{12},n_2)$,

and $D_2 \underset{def}{=} C(l_{22},k_2) \cap C(l_{22},n_2)$.

There exist mappings $\bar{\alpha}$, $\bar{\beta}$, $\bar{\gamma}$ making the following diagram commutative:

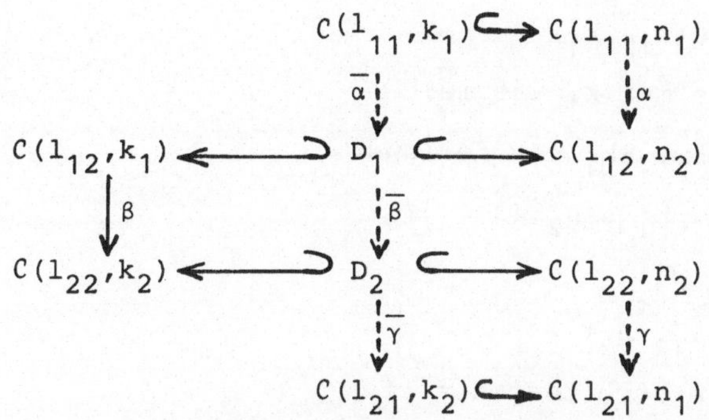

(The arrow $\hookrightarrow$ denotes the inclusion embedding.) $\bar{\alpha}$, $\bar{\beta}$, $\bar{\gamma}$ are uniquely determined by this property and thus are $\sqsubset$-isomorphisms.

(iii) Let $\delta \underset{def}{=} \bar{\gamma} \circ \bar{\beta} \circ \bar{\alpha}$, $\kappa_2 \underset{def}{=} pos(l_{21},k_2)$ and $pos(l_{21},n_1) = pos(l_{11},n_1)$. Then $(l_{11},k_1,\kappa_2,\delta,l_{21},k_2)$ is a substitution.

Proof:

(i) Because n_1 is minimal, we have

$C(l_{11},n_1) = C(l_{11}) \supset C(l_{11},k_1)$ and, analogously,

$C(l_{21},n_1) \supset C(l_{21},k_2)$.

(ii) 1. We will prove: $\alpha[C(l_{11},k_1)] = D_1$.

It holds: $k_1 \;\Delta\; n_1 \Rightarrow \alpha k_1 = k_1$ and $k_1 \;\Delta\; n_2$.

The latter follows from the implications:

$k_1 = \alpha k_1 \in C(l_{12},n_2)$

$k_1 \;\Delta\; n_2$ or $k_1 \sqsupset n_2$

$\alpha k_1 = k_1 \sqsupset n_2 = \alpha n_1 \Rightarrow k_1 \sqsupset n_1$, in contradiction to $k_1 \;\Delta\; n_1$.

1.1. We assume $m \in C(l_{11},k_1)$ and will show $\alpha m \in C(l_{12},k_1)$. Clearly, $\alpha m \in C(l_{12})$.

1.1.1. Consider the case $m \sqsupset k_1$. Hence $\alpha m \sqsupset \alpha k_1 = k_1$ and

$\alpha m \in C(l_{12}, k_1)$.

1.1.2. In case $m \Delta k_1$ assume:

$\exists h \sqsubset \alpha m, \alpha k_1$. It follows:

$h \sqsubset k_1$, $k_1 \Delta n_2$ (see 1.)

$h \Delta n_2$

$h \in C(l_{12}, n_2)$

$\exists i \sqsubset m$ such that $h = \alpha i$

$\alpha i \sqsubset \alpha m, \alpha k_1$

$i \sqsubset m, k_1$ in contradiction to $m \Delta k_1$.

Hence $\alpha m \Delta \alpha k_1 = k_1$ and $\alpha m \in C(l_{12}, k_1)$.

1.2. We assume $s \in D_1$ and will show:

$\exists r \in C(l_{11}, k_1)$ such that $s = \alpha r$.

First note that $s \in C(l_{12}, n_2)$ and thus $\exists r \in C(l_{11}, n_1)$ such that $s = \alpha r$. It remains to show: $r \in C(l_{11}, k_1)$.

Clearly, $r \in C(l_{11})$ and $s \in C(l_{12}, k_1)$ implies $s \sqsupset k_1$ or $s \Delta k_1$.

1.2.1. If $s \sqsupset k_1$, $\alpha r \sqsupset \alpha k_1 = k_1$, hence $r \sqsupset k_1$ and $r \in C(l_{11}, k_1)$.

1.2.2. If $s \Delta k_1$, assume $\exists h$ such that $h \sqsubset r, k_1$. Now

$h \in C(l_{11}) = C(l_{11}, n_1)$ (see (i)) and we infer

$\alpha h \sqsubset \alpha r = s$ and $\alpha h \sqsubset \alpha k_1 = k_1$ in contradiction to $s \Delta k_1$.

Thus $r \Delta k_1$ and $r \in C(l_{11}, k_1)$.

2. and 3. $\beta[D_1] = D_2$ and $\gamma[D_2] = C(l_{21}, k_2)$ are proved analogously.

(iii) We have to show points a) to f) in the definition (2214)(ii).

a) and b) are immediate; c), d), f) follow from (ii); it remains to show e).

Let $m \in C(l_{11})$ and $m \Delta k_1$. We have to derive $\delta m = m$ and $pos(l_{11}, m) = pos(l_{21}, m)$.

Since n_1 is minimal, either $m \Delta n_1$ or $m \sqsupset n_1$ holds.

1. Assume $m \Delta n_1$. It follows, that $m \Delta n_2$ and $m = \alpha m = \beta m = \gamma m$, hence $m = \delta m$. Moreover, $pos(l_{11}, m) = pos(l_{12}, m) = pos(l_{22}, m) =$

$\text{pos}(l_{21}, m)$.

2. Now consider the case $m \sqsupset n_1$ and let $M(m) = \{n_1, m_1, \ldots, m_L\}$. Any m_i is disjoint from n_1 and k_1, the latter because $m \,\Delta\, k_1$. As in 1.1. it can be shown, that $\forall\, i = 1, \ldots, L,\ \delta m_i = m_i$ and $\text{pos}(l_{11}, m_i) = \text{pos}(l_{12}, m_i)$. By assumption, $\text{pos}(l_{11}, n_1) = \text{pos}(l_{12}, n_1)$. Using $\delta n_1 = n_1$ and (ii) we conclude $\delta m \sqsupset n_1, m_1, \ldots, m_L$, hence $\delta m \sqsupset m$ by (2211). In the event of $m \neq \delta m$, we would obtain a countable infinite sequence
$\delta m \underset{\neq}{\sqsupset} m \underset{\neq}{\sqsupset} \delta^{-1} m \underset{\neq}{\sqsupset} \delta^{-2} m \underset{\neq}{\sqsupset} \ldots,$
which is impossible according to (228). Hence $\delta m = m$.
From $\text{pos}(l_{11}, n_1) = \text{pos}(l_{12}, n_1)$ we infer by (2213):
$\text{pos}(l_{11}, m) = \text{pos}(l_{12}, m)$. □

The preceding lemma, which expresses a sort of composability for substitutions of disjoint sub-bodies, can be extended to the case of a finite number of sub-bodies.

(2225) Lemma: Let $k_1, n_i \quad k$, where the n_i are minimal and $n_i \,\Delta\, k_1$ for $i=1,\ldots,L$. Further, the following substitutions are assumed:

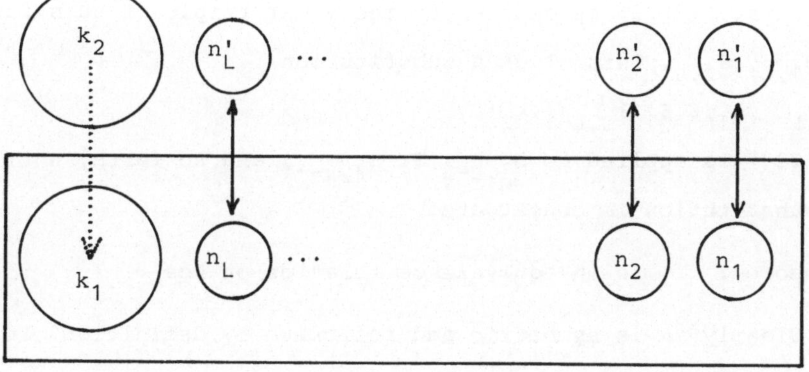

fig. (2226)

$S_{10} = (l_{10}, n_1, \nu_1', \alpha_1, l_{11}, n_1')$

$S_{11} = (l_{11}, n_2, \nu_2', \alpha_2, l_{12}, n_2')$

$\vdots$

$S_{1,i-1} = (l_{1,i-1}, n_i, \nu_i', \alpha_i, l_{1i}, n_i')$

$S_{1,i} = (l_{1,i}, n_{i+1}, \nu_{i+1}', \alpha_{i+1}, l_{1,i+1}, n_{i+1}')$

$\vdots$

$S_{1,L-1} = (l_{1,L-1}, n_L, \nu_L', \alpha_L, l_{1,L}, n_L')$

$S_{1,L} = (l_{1,L}, k_1, \kappa_1, \beta, l_{2,L}, k_2)$

$S_{2,L} = (l_{2,L}, n_L', \nu_L, \gamma_L, l_{2,L-1}, n_L)$

$\vdots$

$S_{2,i+1} = (l_{2,i+1}, n_{i+1}', \nu_{i+1}, \gamma_{i+1}, l_{2,i}, n_{i+1})$

$S_{2,i} = (l_{2i}, n_i', \nu_i, \gamma_i, l_{2,i+1}, n_i)$

$\vdots$

$S_{21} = (l_{21}, n_1', \nu_1, \gamma_1, l_{20}, n_1)$

If $\forall$ i=1,...,L, $\underline{pos}(l_{2,i-1}, n_i) = \underline{pos}(l_{1,i-2}, n_i)$,

$\delta \underset{def}{=} \gamma_1, \ldots, \gamma_L \ \beta \ \alpha_L, \ldots, \alpha_1 | C(l_{1,0}, k_1)$ and

$\kappa_2 \underset{def}{=} pos(l_{2,L}, k_2)$,

then $(l_{1,0}, k_1, \kappa_2, \delta, l_{2,0}, k_2)$ is a substitution.

Proof: First, (2223) is applied to the inner triple of substitutions $S_{1,L-1}$, $S_{1,L}$, $S_{2,L}$. This gives a substitution $T_1 = (l_{1,L-1}, k_1, \kappa, \delta_1, l_{2,L-1}, k_2)$.
Again, (2223) is applied to $S_{1,L-2}, T_1, S_{2,L-1}$, and so forth, until the desired substitution is constructed. □

(2227) Theorem: $\sim$ is an equivalence relation on Pos_φ.

Proof: Clearly, $\sim$ is symmetric and reflexive by definition. It remains to show, that $\pi_1 \sim \pi_2$ and $\pi_2 \sim \pi_3$ implies $\pi_1 \sim \pi_3$.

1. To this end assume $\pi_i = pos(l_i, k_i)$ for i=1,2,3 and

$\{n_i | i=1,\dots,L\} \underset{def}{=} \{n \in B | n \sqsubset l_1, n \Delta k_1, n \sqsubset k_2$ and n is minimal$\}$, which is the set of all minimal sub-bodies of l_1 which hinder the substitution of k_2 for k_1. Further, let $N_1 \underset{def}{=} \{n \in B | n \sqsubset l_1$ or $n \sqsubset l_2$ or $n \sqsubset l_3\}$.

If $\nu_1 \underset{def}{=} pos(l_1, n_1)$, there exists a $(\nu_1', n_1') \in Pos$, such that $\nu_1 \sim \nu_1'$ and $\forall\, n \sqsubset n_1', n \notin N_1$. This follows by (2221) and the finiteness of N_1. $\nu_1 \sim \nu_1'$ implies: $\forall\, \tilde{l}_1 \in \nu_1 \;\exists\, l_{11} \in \nu_1' \;\exists\, \alpha_1$ such that $S_{10} \underset{def}{=} (\tilde{l}_1, n_1, \nu_1', \alpha_1, l_{11}, n_1')$ is a substitution, if the s.c. $\forall n, (n \sqsubset \tilde{l}_1$ and $n \Delta n_1) \Rightarrow n \not\sqsubset n_1'$ is satisfied. It is indeed satisfied for the l_1 and n_1' considered above because

$n \sqsubset l_1 \Rightarrow n \in N_1 \Rightarrow n \not\sqsubset n_1'$.

This procedure will be (L-1)-times repeated, such that the sub-bodies n_i, $i=1,\dots,L$, are consecutively replaced by n_i', where $pos(l_{1,i}, n_{i+1}) = \nu_{i+1} \sim \nu_{i+1}' = pos(l_{1,i+1}, n_{i+1}')$ and $\forall\, n \sqsubset n_i', n \notin N_i \underset{def}{=} N_{i-1} \cup \{n \in B | n \sqsubset n_{i-1}'\}$. The corresponding substitutions will be $S_{1,i} = (l_{1,i}, n_{i+1}, \nu_{i+1}', \alpha_{i+1}, l_{1,i+1}, n_{i+1}')$ for $i=0,\dots,L-1$. Since $k_1 \Delta n_i$ we have $\alpha_i k_1 = k_1$ and $pos(l_{1,i-1}, k_1) = pos(l_{1,i}, k_1)$ (see(2214)(ii) d) and e)).

Hence $pos(l_{1,L}, k_1) = \pi_1$.

2. Claim: $\forall n, (n \sqsubset l_{1,L}$ and $n \Delta k_1) \Rightarrow n \not\sqsubset k_2$.

   Proof: Obviously, it suffices to show this for minimal n. Either $\forall\, i \in \{1\dots L\}$ such that $n \sqsubset n_i'$, or $\forall\, i \in \{1\dots L\}$, $n \Delta n_i'$, because n is minimal. In the first case, $n \sqsubset n_i'$ implies $n \notin N_i \supset N_1$ by the above construction. Thus $n \not\sqsubset l_2$ and $n \not\sqsubset k_2$, which is the assertion. Therefore we can assume $n \Delta n_i'$ for all $i=1\dots L$ and conclude, that n is invariant under all mappings α_i. Thus $n \sqsubset l_{1,L} \Rightarrow n \sqsubset l_1$. Because $n \Delta k_1$, $n \sqsubset k_2$ would imply $n \in \{n_i | i=1\dots L\}$ in contradiction to $n \sqsubset l_{1,L}$, since $l_{1,L}$ contains no n_i by its very construction. This completes the proof of the above claim, which is just the s.c. for $\pi_1 \sim \pi_2$.

3. This proves the existence of a substitution

   $S'_{1,L} = (l_{1,L}, k_1, \pi_2, \beta_2, l'_{2,L}, k_2)$.

   The construction also insures, that the s.c.

   $\forall n, (n \sqsubset l_1$ and $n \Delta k_1) \Rightarrow n \not\sqsubset k_3$, which we may assume, implies the modified s.c.

   $\forall n, (n \sqsubset l'_{2,L}$ and $n \Delta k_2) \Rightarrow n \not\sqsubset k_3$. Hence $\pi_2 \sim \pi_3$ yields a substitution

   $S'_{3,L} = (l'_{2,L}, k_2, \pi_3, \beta_3, l_{3,L}, k_3)$.

4. Set $S_{1,L} = S'_{3,L} \circ S'_{1,L}$ and construct substitutions

   $S_{3,i} = l_{3,i}, n'_i, \nu_i, \gamma_i, l_{3,i-1}, n_i)$ for i=L...1. This is possible by dint of $\nu_i \sim \nu'_i$, if the corresponding s.c. are satisfied, which simply reduce to $n_i \not\sqsubset l_{3,i}$. Since the n_i are minimal, there are only 4 possibilities for $n_i \sqsubset l_{3,i}$:

4.1. $n_i \sqsubset n_k$ for some $k > i$, which is impossible since n_k is minimal.

4.2. $n_i \sqsubset n'_j$ for some $j \leq i$ is also impossible, since this would imply by construction $n_i \notin N_j \supset N_1$, however $n_1 \in N_1$.

4.3. $n_i \sqsubset k_3$. This would violate the above s.c. for $\pi_2 \sim \pi_3$.

4.4. $n_i \,\Delta\, n_k, n'_j, k_3$ for all $k > i,j$. In this case n_i is invariant under all γ_k, $k > i$ and β_3. Hence $n_i \,\Delta\, k_2$ is in contradiction to $n_i \sqsubset k_2$.

5. We have $pos(l_{1,i-1}, n_i) = pos(l_{2,i-1}, n_i)$ and can apply (2225) to the sequence $S_{10} \ldots S_{1L}, S_{2L}, \ldots S_{21}$. This yields a substitution $(l_1, k_1, \pi_3, \delta, l_3, k_3)$ and thereby proves $\pi_1 \sim \pi_3$. □

In sequel, we can assume that the s.c. is always satisfied since we need only apply the technique used in the previous proof.

(2228) Definition: $R_{(\varphi)} \underset{def}{=} Pos_{(\varphi)}/\sim$ will be called the set of <u>spatial regions</u>. The canonical surjection $\nu_\sim : Pos \to R$ will be written as $\alpha \to reg(\alpha)$ and, with an abuse of language, we will also write $reg(pos(l,k),k) = reg(l,k)$ for $l \in B_k \cap \varphi$

(2229) Definition: Let $A, B \in \mathcal{R}$.

$A \prec B \underset{\text{def}}{\Leftrightarrow} \exists l, k_A, k_B \in B$ such that $A = reg(l,k_A), B = reg(l,k_B)$ and $k_A \sqsubset k_B$.

(2230) <u>Axiom B4:</u> Let $l \in B_{k_1} \cap B_{k_2} \cap \varphi$ such that $k_1 \Delta k_2$. Then $reg(l,k_1) \neq reg(l,k_2)$.

(2231) Lemma: Let $l \in B_{k_2}$ and $k_1 \underset{\neq}{\sqsubset} k_2$. Then

$A_1 \underset{\text{def}}{=} reg(l,k_1) \underset{\neq}{\prec} reg(l,k_2) \underset{\text{def}}{=} A_2$.

Proof: Clearly $A_1 \prec A_2$ by (2229). Now assume $A_1 = A_2$. By (2212), $M(k_1) \underset{\neq}{\subset} M(k_2)$, say $m \in M(k_2)$, but $m \notin M(k_1)$. Replace m in l by an equivalent m' (see (2221)), thus obtaining $l' \in B_{k_2'}$ and $pos(l',k_1) \sim pos(l,k_2) \underset{\text{def}}{=} \pi$. Let $(l',k_1,\pi,\alpha,l'',k_2)$ be the corresponding substitution. From $m' \Delta k_1$ it follows $\alpha m' = m'$ and $pos(l',m') = pos(l'',m')$ (see (2214)(ii)e)). Further, $reg(l,m) = reg(l',m')$ by construction and $pos(l,k_2) = \pi = pos(l'',k_2)$, hence $pos(l,m) = pos(l'',m)$. We conclude: $reg(l'',m) = reg(l'',m')$ in contradiction to $m \Delta m'$ and (2230). □

(2232) Proposition: $\prec$ is a partial ordering on $\mathcal{R}$.

Proof: Clearly, $\prec$ is reflexive. Now assume $A \prec B$ and $B \prec A$, i.e. $A = reg(l,k_A) = reg(m,n_A)$, $B = reg(l,k_B) = reg(m,n_B)$, $k_A \sqsubset k_B$ and $n_B \sqsubset n_A$. By the existence of a substitution $(l,k_B, pos(m,n_B),\alpha,l',n_B)$, hence $A = reg(l',\alpha k_A) = reg(l',n_A)$ and $n_A \sqsubset n_B = \alpha k_B \sqsubset \alpha k_A$ we infer from (2231): $n_A = \alpha k_A$, $n_A = n_B$, $A = B$.

Next assume $A \prec B \prec C$, i.e.:

$A = reg(l,k_A)$

$B = reg(l,k_B) = reg(m,n_B)$

$C = reg(m,n_C)$ and $k_A \sqsubset k_B$, $n_B \sqsubset n_C$.

As above, we consider the substitution $(m,n_B,pos(l,k_B),\alpha,m',k_B)$. It follows from (2222), that $pos(m',\alpha n_C) \sim pos(m,n_C)$. Hence $k_A \sqsubset k_B = \alpha n_B \sqsubset \alpha n_C$ implies $A \prec C$. □

(2233) Definition: Let $A, B \in R$

(i) $A\ m\ B \underset{def}{\Leftrightarrow} \exists\ C \in R$ such that $C \prec A,B$, ("A <u>meets</u> B").

(ii) $A \vartriangle B \underset{def}{\Leftrightarrow}$ not $(A\ m\ B)$, ("A and B are disjoint").

(2234) <u>Axiom B5:</u> Let $A, B \in R$.

(i) $\exists\ D \in R$ such that $A, B \prec D$,

(ii) $A\ m\ B \Rightarrow$ the $\prec$-supremum $A \vee B$ exists,

(iii) $\{C \in R \mid C \prec A,B\}$ contains a finite number of $\prec$-maximal spatial regions,

(iv) $\forall\ C \in R\ \exists\ D \in R$ such that $C \vartriangle D$.

(v) $\forall\ C \in R, (AmB$ and $A,B \vartriangle C) \Rightarrow (A \vee B) \vartriangle C$.

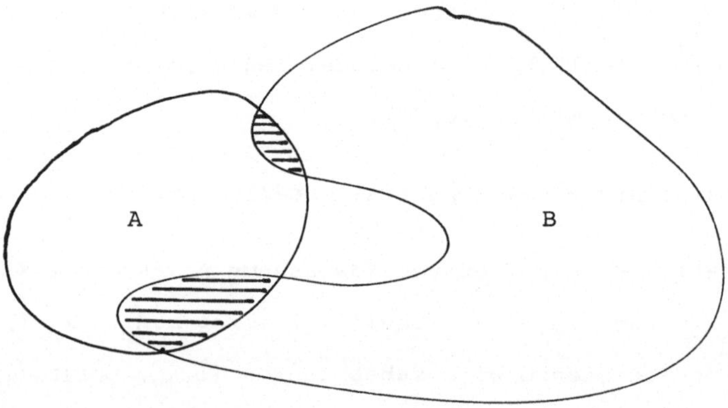

fig. (2235)

Intuitively, spatial regions are connected, since they are represented by rigid bodies. Hence, we do not postulate that the infimum, $A \wedge B$ is a spatial region, for it can indeed be disconnected.

The postulates of (2234) basically mean that one can produce and fixate a sufficient variety of rigid bodies. Although, it would be very complicated to express these laws on the $(B, \sqsubset, R)$-level. Mathematically, (2234) enables us to embed $\mathcal{R}$ into a weakly distributive lattice R consisting of finite, disjoint unions of spatial regions.

(2236) Lemma: Let $C, A_i \in \mathcal{R}$, $i=1...n$, and $\forall\, i=1...n$, $C \prec A_i$. Then the supremum

$$A = \bigvee_{i=1...n} A_i \in \mathcal{R} \text{ exists.}$$

Proof (by induction on n): The assertion is trivial for $n = 1$. Assume: $\forall\, i=1...k$, $(k<n)$ $C \prec A_i$ and $B = \bigvee_{i=1...k} A_i \in \mathcal{R}$. We have $C \prec B, A_{k+1}$ and $A \underset{def}{=} B \vee A_{k+1} \in \mathcal{R}$ by (2234)(ii). Let us show that A is the supremum of all A_i, $i=1...k+1$. Clearly, $A \succ A_i$ for $i=1...k+1$. Let $D \succ A_i$ for $i=1...k+1$. We conclude $D \succ B$ and, from $D \succ A_{k+1}$, $D \succ B \vee A_{k+1} = A$. □

(2237) Definition: Let a be any subset of $\mathcal{R}$ and ∞ the equivalence relation on a generated by m.

Therefore, $|a/\infty| = 1$ is equivalent to:

$\forall\, A, B \in a\ \exists\, n \in \mathbb{N}\ \exists\, A_1, ... A_n \in a$ such that $A_1 = A$, $A_n = B$ and $(\forall\, i=1...n-1, A_i\, m\, A_{i+1})$.

(2238) Lemma: Let $a \subset \mathcal{R}$ be finite and $|a/\infty| = 1$. Then $\bigvee_{A \in a} A \in \mathcal{R}$ exists.

Proof (by induction on $|a|$): If a is empty or a singleton set, the assertion is trivial. Now assume, that it holds for $|\tilde{a}| \leq k$. Let $|a| = k + 1$, $A \in a$ and consider $a' = a \setminus \{A\}$. a' is the finite, disjoint union of its ∞-equivalence classes, which have less than k + 1 elements. Say $|a'/\infty| = n$ and consider $\alpha_i \in a'/\infty$ where $|\alpha_i| \leq k$. Set $B_i = \bigvee_{D \in \alpha_i} D$ for $i=1...n$ and $A_i = B_i \vee A$. B_i exists by induction hypothesis and A_i by $B_i\ m\ A$.

Now, $C = \bigvee_{i=1\ldots n} A_i$ exists by (2236) and is easily shown to be the supremum $\bigvee_{D\in a} a$. □

(2239) Lemma: Let $a, b \subset \mathcal{R}$ be finite and $|a/\infty| = |b/\infty| = 1$, further $\forall A \in a \ \forall B \in b,\ A \,\Delta\, B$.

Then $(\bigvee_{A\in a} A) \,\Delta\, (\bigvee_{B\in b} B)$ holds.

Proof:

1. Let $B \in b$. We will prove $B \,\Delta\, (\bigvee_{A\in a} A)$ by induction on $k = |a|$.

1.1. For $k = 2$ the assertion is just (2234)(v).

1.2. Assume (2239) for all $\tilde{a}$ such that $|\tilde{a}| \leq k$. Consider $|a| = k + 1$ and write $\bigvee_{A\in a} A = \bigvee_{i=1\ldots n} A_i$, $A_i = (\bigvee_{D\in\alpha_i} D) \vee A$ as in the proof of (2238). Because $|\alpha_i| \leq k - 1$ and $n \leq k$, the induction hypothesis can be applied consecutively in order to derive the result.

2. We have proved:
$(\forall\tilde{A}\in\tilde{a},\ \tilde{A}\Delta\tilde{B}) \Rightarrow \tilde{B} \,\Delta \bigvee_{\tilde{A}\in\tilde{a}} \tilde{A}$. Now set $\tilde{B} = \bigvee_{A\in a} A$, $\tilde{a} = b$ and use $(\forall B\in b,\ B\Delta \bigvee_{A\in a} A)$, which has been shown in 1., to conclude:
$(\bigvee_{A\in a} A) \,\Delta\, (\bigvee_{B\in b} B)$. □

(2240) Definition:

(i) $R \underset{def}{=} \{a\subset\mathcal{R} \mid a \text{ is finite and } \forall A,B\in a, A=B \text{ or } A\Delta B\}$. Note: $\emptyset \in R$. The elements $a \in R$ will be called <u>regions</u>.

(ii) Let $a, b \in R$. $a < b \underset{def}{\Leftrightarrow} \forall A \in a \ \exists B \in b$ such that $A \prec B$.

(2241) Proposition: R is partially ordered by $<$.

Proof: Reflexivity and transitivity are immediate consequences of the corresponding properties of $(\mathcal{R},\prec)$. In order to show antisymmetry, assume $a < b$ and $b < a$, i.e.:

$\forall A \in a \; \exists B \in b$ such that $A \prec B$, and $\forall B \in b \; \exists A' \in a$ such that $B \prec A'$. Hence there are two mappings $\beta : a \to b$ and $\gamma : b \to a$ such that $\forall A \in a$, $A \prec \gamma\beta A$. Thus $A \,\Delta\, \gamma\beta A$ is impossible and $A = \gamma\beta A$ by definition of R. Now $A \prec \beta A \prec \gamma\beta A = A$ and, by antisymmetry of $\prec$, $\beta A = A = \gamma A$. The above assertions thus imply $a \subset b$ and $b \subset a$, hence $a = b$. □

(2242) Theorem: For any 2 regions $a, b \in R$, the supremum $a \vee b$ and the infimum $a \wedge b$ exist w.r. to the parital ordering $<$.

Proof:

1. Definition of $a \vee b$.

Consider the equivalence relation ∞ on $a \cup b$ and any $\alpha \in (a \cup b)/\infty$. By (2238), $c(\alpha) = \bigvee_{A \in \alpha} A \in \mathcal{R}$ exists. We define $a \vee b \underset{def}{=} \{c(\alpha) \mid \alpha \in (a \cup b)/\infty\}$.

1.1. $a \vee b \in R$.

We have to show that $a \vee b$ consists of a finite number of disjoint spatial regions. By definition of $a \vee b$, this means, that $c(\alpha) \,\Delta\, c(\beta)$, whenever $\alpha, \beta \in (a \cup b)/\infty$ are different, i.e. $\forall A \in \alpha \; \forall B \in \beta$, $A \,\Delta\, B$. But this is just the assertion of (2239).

1.2. $a < (a \vee b)$.

Assume $A \in a$. It follows:

$A \in a \cup b$

$\exists \alpha$ such that $A \in \alpha \in (a \cup b)/\infty$

$A \prec c(\alpha) = \bigvee_{C \in \alpha} C \in a \vee b$.

1.3. Equivalently, $b < (a \vee b)$.

1.4. $a, b < d \Rightarrow (a \vee b) < d$.

Assume $c(\alpha) \in a \vee b$, $\alpha \in (a \cup b)/\infty$.

1.4.1. Let $A, B \in \alpha$ and $A \in a$, $B \in b$ $A\ m\ B$. Since $a < d$, $\exists\ D_A \in d$ such that $A \prec D_A$, analogously: $\exists\ D_B \in d$ such that $B \prec D_B$. Now $A\ m\ B \Rightarrow D_A\ m\ D_B$, and, since $d \in R$, $D_A = D_B$.

1.4.2. Let $F, G \in \alpha$, i.e.: there exist H_i, $i=1\ldots n$, such that $F = H_1$, $G = H_n$ and $H_i\ m\ H_{i+1}$ for $i=1\ldots n-1$. Thus H_i and H_{i+1} cannot be both elements of a, or of b. Assume for instance: $H_i \in a$ and $H_{i+1} \in b$. 1.4.1 shows: $\exists\ D \in d$ such that $H_i, H_{i+1} \prec D$. D is unique, since the spatial regions in d are disjoint. By induction on i we infer $F, G \prec D$.

1.4.3. From 1.4.2. we conclude

$$c(\alpha) = \left(\bigvee_{F\in\alpha} F\right) \prec D \in d, \text{ hence } (a\vee b) < d.$$

2. Definition of $a \wedge b$.

2.1. First we define for $A, B \in \mathcal{R}$:

$A \curlywedge B \underset{\text{def}}{=}$ the set of maximal spatial regions of $\{C\in\mathcal{R} | C\ A,B\}$. $A \curlywedge B$ is finite by (2234)(iii).

Let $C_1, C_2 \in A \curlywedge B$ and assume $C_1\ m\ C_2$. By (2234)(ii), $C_1 \curlyvee C_2 \in \mathcal{R}$ exists and satisfies $(C_1 \curlyvee C_2) \prec A, B$. By maximality of C_1, C_2, we have $C_1 = C_1 \curlyvee C_2 = C_2$.

This proves $A \curlywedge B \in R$.

2.2. Now $a \wedge b \underset{\text{def}}{=} \bigcup_{A\in a, B\in b} A \curlywedge B$ is a finite subset of $\mathcal{R}$. Let $C_1, C_2 \in a \wedge b$. That means $C_i \in A_i \curlywedge B_i$, $A_i \in a$, $B_i \in b$ for $i=1,2$. If $A_1 = A_2$ and $B_1 = B_2$, 2.1. shows: $C_1 = C_2$ or $C_1 \vartriangle C_2$. Assume for instance $A_1 \vartriangle A_2$, then $C_i \prec A_i$ implies $C_1 \vartriangle C_2$. At any case: $a \wedge b \in R$.

2.3. $(a\wedge b) < a, b$ is clear by definition.

2.4. $c < a, b \Rightarrow c < a \wedge b$.

$c < a, b$ means: $\forall\ C \in c\ \exists\ A \in a, B \in b$ such that $C \prec A, B$.

Then there exists a maximal $M \in A \curlywedge B$ such that $C \prec M$.

$M \in a \wedge b$ shows $c < a \wedge b$. □

(2243) Proposition: The mapping $i : \mathcal{R} \to R$, $i(A) \underset{def}{=} \{A\}$, is an order-preserving injection.

Proof: immediate. □

(2244) Definition: A lattice $(R,\wedge,\vee)$ with smallest element 0 will be called <u>weakly distributive</u>, iff

$\forall\, a, b, c \in R,\ (a\wedge c=0 \text{ and } b\wedge c=0) \Rightarrow (a\vee b) \wedge c = 0.$

By induction one easily proves the

(2245) Lemma: Let $(R,\wedge,\vee,0)$ be a weakly distributive, $n \in \mathbb{N}$. Then $(\forall i=1\ldots n,\ a_i\wedge c=0) \Rightarrow \left(\bigvee_{i=1}^{n} a_i\right) \wedge c = 0.$

Each distributive lattice is weakly distributive, but not conversely (see counterexample (329)).

(2246) Theorem: $(R,\wedge,\vee,\emptyset)$ is a weakly distributive lattice. $R \neq \{\emptyset\}$.

Proof: By (2241) and (2242) it remains to show the weak distributivity.

Thus assume $a \wedge c = b \wedge c = \emptyset$. It is enough to show:

$\forall\, C \in c\ \forall\, D \in a \vee b,\ C \curlywedge D = \emptyset.$

D is of the form $D = \bigvee_{D_i \in \alpha} D_i$, $\alpha \in (a\cup b)/\infty$. Each spatial region D_i is an element of a or of b, hence $C \curlywedge D_i = \emptyset$. Analogously to (2245) it follows from (2234)(v) by induction, that $C \curlywedge (\bigvee_{i=1\ldots n} D_i) = C \curlywedge D = \emptyset$.

$R \neq \{\emptyset\}$ follows, if we anticipate Axiom $J1$ (2311), which implies $F \neq \emptyset$, $B_R \neq \emptyset$, $\mathcal{R} \neq \emptyset$. □

2.3 CONSTRUCTION OF TRANSPORT MAPPINGS

According to the ideas developed in section 2.1 we have to define the transport mapping induced by a pair of positions of the same body. First we will state 3 additional axioms concerning configurations. We recall that $C(l) = \{m \in B \mid l \in B_m\}$ by def. (2214)(i).

(231) <u>Axiom B6:</u> $\forall\ l \in B_R\ \forall\ D \subset C(l)\quad \exists\ m \in B_R$ such that $D = C(m)$ and $\forall\ k \in D$, $pos(l,k) = pos(m,k)$.

This means simply, that those parts of l which are not contained in D could be either omitted or replaced by parts of reference bodies.

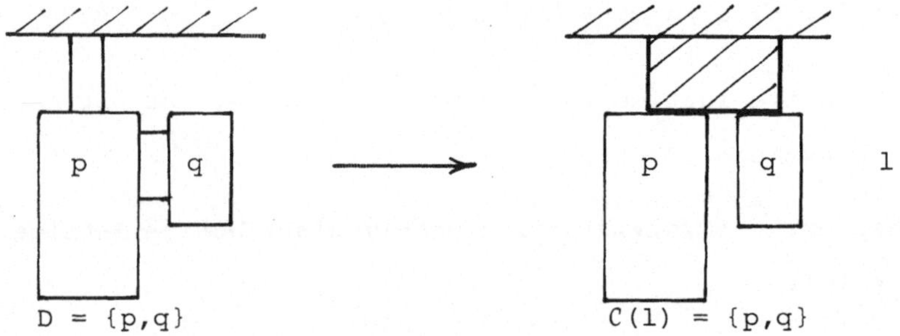

fig. (232)

(233) <u>Axiom B7:</u> Let $k_1, k_2 \in B$, $l, m \in B_{k_1} \cap B_{k_2}$ be such that $pos(l,k_i) = pos(m,k_i)$ for $i=1,2$. Then
$\exists\ M \in \mathbb{N}\ \exists\ l_1, \dots l_M \in B_{k_1} \cap B_{k_2}$ such that $l = l_1$, $m = l_M$ and
$\forall\ \nu \in \{1, \dots M-1\}\ l_\nu \sqsubset l_{\nu+1}$ or $l_\nu \sqsupset l_{\nu+1}$.

By the definition of $pos(l,k_i) = pos(m,k_i)$ there exist such sequences of configurations from B_{k_1} and B_{k_2} separately. Axiom B7 postulates the existence of a "joint" sequence.

(234) Definition: A configuration $l \in B_R$ will be called <u>normal</u> iff $C(l)$ possesses a $\sqsubset$-greatest element $\hat{l}$. B_{RN} is the set of all normal configurations.

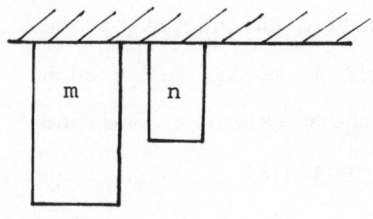

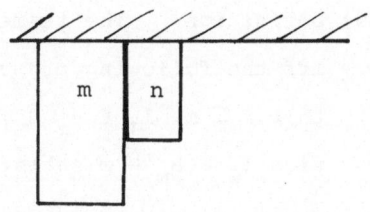

$l_1 \notin B_{RN}$ $l_2 \in B_{RN}$, $\hat{l}_2 = mn$

fig. (235)

(236) Definition: Let $l, m \in B_R$. $l \; ¢ \; m \underset{def}{\Leftrightarrow} C(l) \subset C(m)$ and $\forall\, k \in C(l)$, $pos(l,k) = pos(m,k)$.

(237) <u>Axiom B8:</u> Let $\bar{l}_\nu \in B_R$ for $\nu=1\ldots N$ and $\bar{l}_1\ I_1\ \bar{l}_2\ I_2\ \bar{l}_3 \ldots \bar{l}_N$, $I_\nu = \sqsubset$ or $I_\nu = \sqsupset$. Then $\exists\ l_\nu \in B_{RN}$, $\nu=1\ldots N$, such that $\bar{l}_\nu \; ¢ \; l_\nu$ and $\hat{l}_1\ I_1\ \hat{l}_2\ I_2\ \hat{l}_3 \ldots \hat{l}_N$.

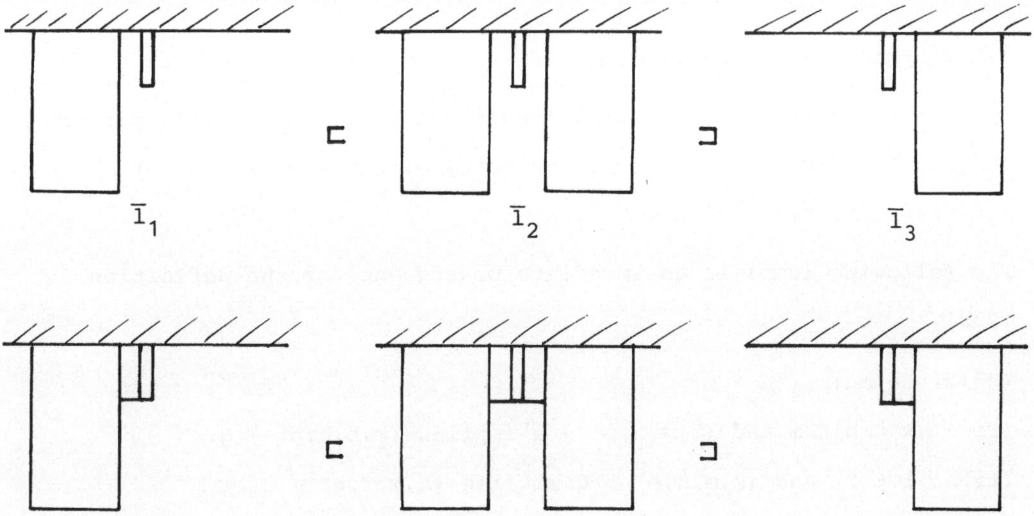

$\bar{l}_1$ $\bar{l}_2$ $\bar{l}_3$

l_1 l_2 l_3

fig. (238)

(239) Definition: The frame $\varphi \in F$ will be called an <u>inertial frame</u>, iff the following property holds: For all k, m, l_1, n $\in$ B such that $k \sqsubset m$, $l_1 \in B_m \cap \varphi$ and $n \in B_k \cap \varphi$, there exists a configuration $l_2 \in B_m \cap \varphi$ satisfying $pos(l_2,k) = pos(n,k)$.

In this case we will write $(k,m,l_1,n) \rightarrow l_2$.

The subset of inertial frames will be denoted by $J \subset F$.

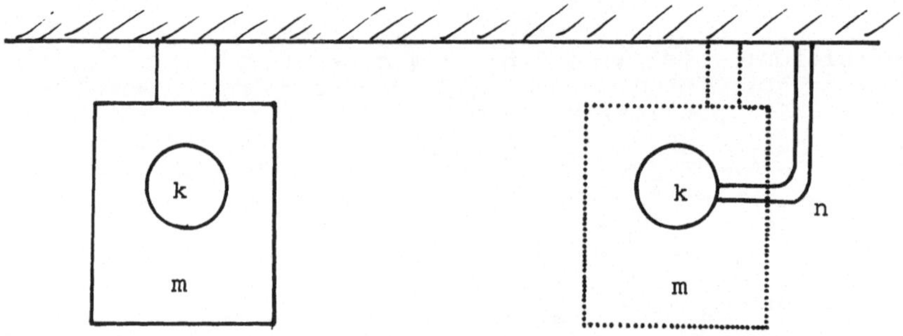

l_1 l_2

fig. (2310)

(2311) <u>Axiom J1:</u> $J \neq \emptyset$.

Thus we postulate that inertial frames can be found and will perform the further developement of pre-geometry within some fixed but arbitrary inertial frame $\varphi \in J$.

The following lemma is an immediate consequence of the definition (239).

(2312) Lemma:

(i) $k \sqsubset n \sqsubset m$ and $(k,m,h,p) \rightarrow q$ implies $(k,n,h,p) \rightarrow q$.

(ii) $l \in B_m$ and $(k,m,h,p) \rightarrow q$ implies $(k,m,l,p) \rightarrow q$

(iii) $(k_2,m,h_1,q) \rightarrow q_1$, $q \in B_m$ and $k_1 \sqsubset m$ implies $pos(q,k_1) = pos(q_1,k_1)$.

(Use axiom B2.)

(iv) $(k,m,h,p) \to q$ and $pos(p,k) = pos(\bar{p},k)$ implies $(k,m,h,\bar{p}) \to q$.

(v) $h \sqsubset k$ and $(k,m,l,p) \to q$ implies $(h,m,l,p) \to q$.

(2313) Definition:

(i) A (positional) <u>N-chain</u> is a sequence of positions $\sigma = (\pi_1,k_1;\pi_2,k_2;\ldots\pi_N,k_N)$, $(\pi_\nu,k_\nu) \in Pos_\varphi$, such that for $\nu=1\ldots N-1$, $(\pi_\nu,k_\nu)\ I_\nu\ (\pi_{\nu+1},k_{\nu+1})$, where I_ν is either $\sqsubset$ or $\sqsupset$ (see (2210)(vi)). Without danger of confusion we may write: $\sigma \equiv (\pi_1,k_1)\ I_1\ (\pi_2,k_2)\ I_2\ \ldots\ (\pi_N,k_N)$.

(ii) σ is called <u>cyclic</u> iff $(\pi_1,k_1) = (\pi_N,k_N)$.

(iii) Two chains σ, τ are called <u>congruent</u>, $\sigma \,\square\, \tau$, iff $\sigma \equiv (\pi_1,k_1)\ I_1\ (\pi_2,k_2)\ I_2\ \ldots\ (\pi_N,k_N)$ and $\tau \equiv (\rho_1,k_1)\ I_1\ (\rho_2,k_2)\ I_2\ \ldots\ (\rho_N,k_N)$.

(iv) In particular, two positions (π_1,k_1), (π_2,k_2) are congruent iff $k_1 = k_2$.

(v) An N-chain $\sigma = (\pi_1,k_1)\ I_1\ \ldots\ (\pi_N,k_N)$ is called <u>weakly transportable</u> iff $\forall\ \nu \in \{1\ldots N\}\ \forall\ (\rho,k_\nu) \in Pos_\varphi$ there exists an N-chain τ which contains (ρ,k_ν) and satisfies $\sigma \,\square\, \tau$. Clearly, τ is unique (using axiom B3).

(vi) A cyclic N-chain σ is called <u>transportable</u> iff it is weakly transportable and each τ satisfying (v) will be cyclic.

Thus the defining property of inertial frames (239) is just: Any 2-chain is weakly transportable. This can easily be generalized:

(2214) Proposition: Any N-chain is weakly transportable.

Proof: By induction. $\square$

(2315) Proposition: Any two positions can be joined by a 5-chain. That is: $\forall\ (\pi_1,k_1), (\pi_5,k_5) \in Pos_\varphi$ there exists a 5-chain $\sigma \equiv (\pi_1,k_1)\ I_1\ (\pi_2,k_2)\ \ldots\ (\pi_5,k_5)$.

Proof: By (2234)(i) there exists a region $D \in R$ such that $A_i \underset{def}{=} reg(\pi_i, k_i) < D$ for i=1,2 and, by (2234)(iv), a region $C \in R$ such that $C \Delta D$, hence $C \Delta A_i$; and further there exist regions $B_1, B_2 \in R$ satisfying $B_i > A_i, C$ for i=1,2. After appropriate substitutions the sequence of regions

$A_1 < B_1 > C < B_2 > A_2$

may be represented by some sequence of positions

$(\pi_1, k_1) \sqsubset (\beta_1, b_1) \sqsupset (\gamma, c) \sqsubset (\beta_2, b_2) \sqsupset (\pi_2, k_2)$.

This is the required 5-chain. □

The crucial point in the theory of transport is the transportability of cyclic chains. The following lemma will imply the transportability of cyclic 4-chains.

(2216) Lemma: Consider bodies $k_i \sqsubset s_2 \sqsubset s_1$, for i=1,2, $h_i \in B_{s_i}$ and $(k_1, s_i, h_i, p) \to q_i$. Then $pos(q_1, k_2) = pos(q_2, k_2)$.

Proof: By assumption, $(k_1, s_1, h_1, p) \to q_1$,
(2312)(i) implies $(k_1, s_2, h_1, p) \to q_1$,
(2312)(ii) implies $(k_1, s_2, h_2, p) \to q_1$.
By assumption, $(k_1, s_2, h_2, p) \to q_2$.
Hence $pos(q_1, k_1) = pos(p, k_1) = pos(q_2, k_1)$. Now axiom B2 (2213) yields $pos(q_1, s_2) = pos(q_2, s_2)$, hence $pos(q_1; k_2) = pos(q_2, k_2)$. □

(2317) Lemma: Let $k_i \sqsubset m, n \in B$ (i=1,2) $h \in B_m$, $l \in B_n$ $pos(h, k_i) = pos(l, k_i)$, $(k_1, m, h, p) \to q$, $(k_1, n, l, p) \to r$. Then it follows that $pos(q, k_2) = pos(r, k_2)$.

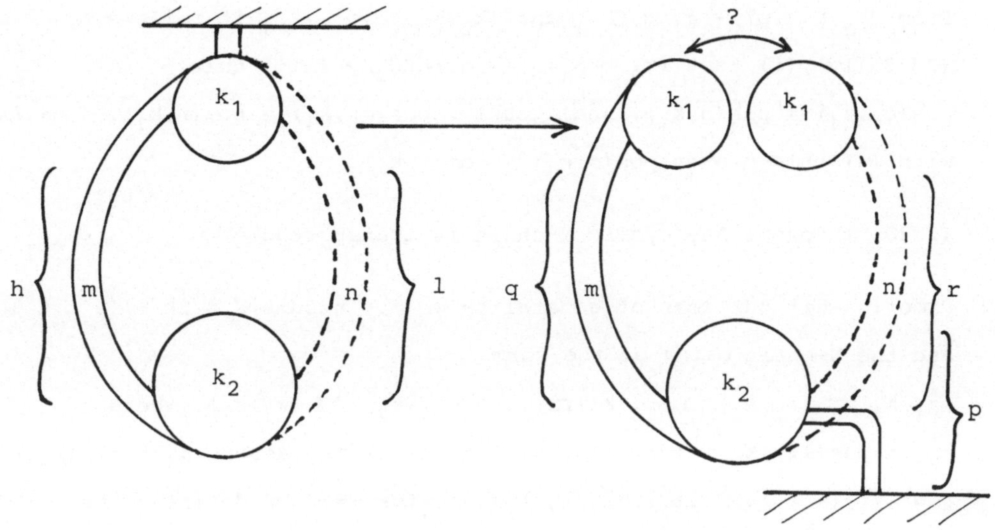

fig. (2318)

(2319) Corollary: Any cyclic 5-chain is transportable.

Proof: Without loss of generality we may consider a 5-chain of the form $(\pi_1,k_1) \sqsubset (\pi_2,m) \sqsupset (\pi_3,k_2) \sqsubset (\pi_4,n) \sqsupset (\pi_1,k_1)$.
Let $h \in B_m$, $l \in B_n$, $\pi_2 = pos(h,m)$, $\pi_4 = pos(l,n)$ and (without loss of generality) $p \in B_{k_2}$ be the position determining the transport: $(k_2,m,h,p) \to q$, $(k_2,n,l,p) \to r$. Thus we have to prove: $pos(q,k_1) = pos(r,k_1)$, which is just the claim of (2317). □

Proof of (2317):
By virtue of $pos(h,k_i) = pos(l,k_i)$ and axiom B7 (233) there exists a sequence of configurations $\bar{h}_\nu \in B_{k_1} \cap B_{k_2}$, $\nu=1...L$, such that $h = \bar{h}_1\ I_1\ \bar{h}_2\ I_2\ ...\ I_{L-1}\ \bar{h}_L = l$ ($I_\nu = \sqsubset$ or $\sqsupset$). Let $h_\nu \in B_{RN}$ be the corresponding sequence of normal configurations according to axiom B8 (237) and s_ν the greatest element of $C(h_\nu)$ (see (234)). We put $(k_2,s_\nu,h_\nu,p) \to q_\nu$ for $\nu=1...L$. We infer: $k_1,k_2 \in C(\bar{h}_\nu)$, $\bar{h}_\nu \not\subset h_\nu$ $\Rightarrow k_1,k_2 \in C(h_\nu) \Rightarrow k_1,k_2 \sqsubset s_\nu$. Since $s_\nu\ I_\nu\ s_{\nu+1}$ (238) we may apply lemma (2316) and conclude $pos(q_\nu,k_1) = pos(q_{\nu+1},k_1)$. By induction, (*) $pos(q_1,k_1) = pos(q_L,k_1)$.

From $(k_2,m,h,p) \to q$, $m \sqsubset s_1$ and $(k_2,s_1,h_1,p) \to q_1$ it follows by (2312)(iv),(i) that $(k_2,m,h_1,q) \to q_1$ and by (iii) that $pos(q_1,k_1) = pos(q,k_1)$. Analogously: $pos(q_L,k_1) = pos(r,k_1)$. Together with (*) this proves $pos(q,k_1) = pos(r,k_1)$. □

(2320) Theorem: Any cyclic N-chain is transportable.

Proof: Without loss of generality we may assume $N = 2L + 1$, $L \in \mathbb{N}$, and the N-chain being ot the form
$(\pi_1,k_1) \sqsubset (\pi_2,k_2) \sqsupset (\pi_3,k_3) \sqsubset \ldots (\pi_{2L},k_{2L}) \sqsupset (\pi_1,k_1)$, where $\pi_{2i} = pos(l_{2i},k_{2i})$ for $i=1\ldots L$. Further we may assume $\rho_1 = pos(n_o,k_1)$ and (*) $(k_{2i-1},k_{2i},l_{2i},n_{2(i-1)}) \to n_{2i}$ for $i=1\ldots L$ defining the congruent chain. We have to show that it is cyclic, i.e. $pos(n_o,k_1) = pos(n_{2L},k_1)$.

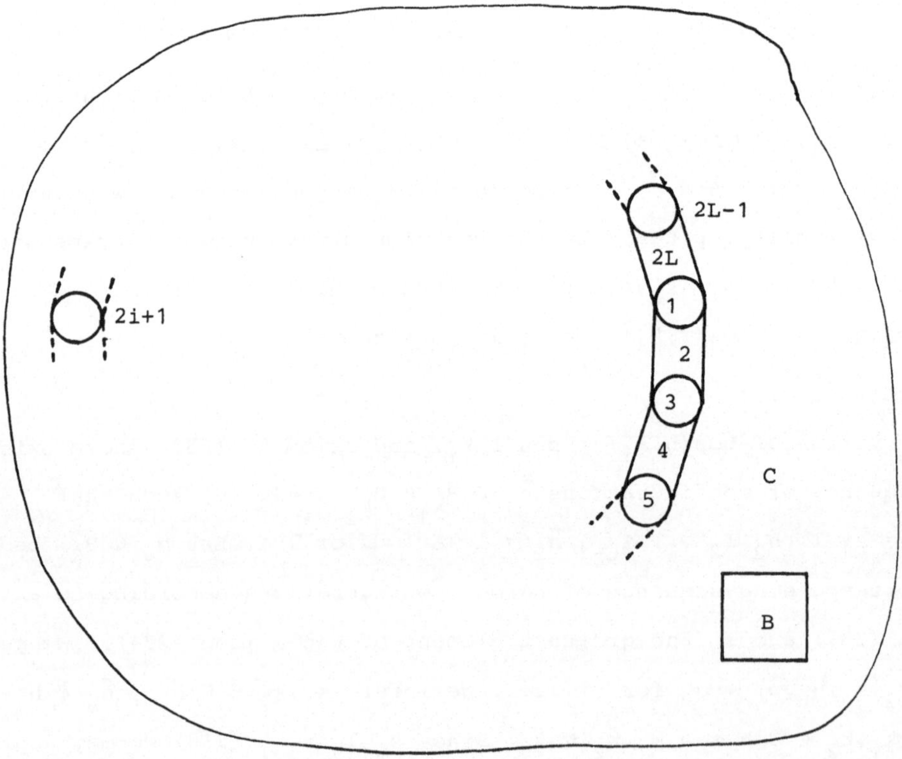

fig. (2321)

Set $A = \bigvee_{i=1...L} reg(\pi_{2i},k_{2i})$, $B \Delta A$ and $C \succ A, B$ (see axiom B5 (2234)). After some appropriate substition we will obtain bodies $b \sqsubset c_i \in B$ together with configurations $m_{2i} \in B_R$ (i=1...L) such that

$\pi_{2i} = pos(m_{2i},k_{2i}) \sqsubset pos(m_{2i},c_{2i})$, $reg(m_{2i},c_{2i}) = C$, $reg(m_{2i},b) = B$ and $pos(m_{2i},b) = \beta$ for some constant position β. Now let

(**) $(k_{2i},c_{2i},m_{2i},n_{2i}) \to p_{2i}$

for i=1...L. We claim:

(2320a) $pos(p_2,b) = pos(p_4,b)$.

Proof: From $(k_2,c_2,m_2,n_2) \to p_2$ and (2312)(v) follows $(k_3,c_2,m_2,n_2) \to p_2$. Analogously: $(k_3,c_4,m_4,n_4) \to p_4$ and by dint of $pos(n_4,k_4) = pos(n_2,k_3)$ and (2312)(iv): $(k_3,c_4,m_4,n_2) \to p_4$. Now we may state the assumptions of lemma (2317) in the following form:

Let $k_N \underset{def}{=} b$. Then

$k_i \sqsubset c_2, c_4$ (i=3,N), $m_2 \in B_{c_2}$, $m_4 \in B_{c_4}$, $pos(m_2,k_i) = pos(m_4,k_i)$,

$(k_3,c_2,m_2,n_2) \to p_2$ and $(k_3,c_4,m_4,n_2) \to p_4$.

Hence (2317) yields: $pos(p_2,k_N) = pos(p_4,k_N)$. □

By the same method

$pos(p_2,b) = pos(p_4,b) = ... = pos(p_{2L},b)$ is proved, hence

(***) $(b,c_{2L},m_{2L},p_2) \to p_{2L}$.

From $(\pi_1,k_1) \sqsubset (\pi_2,k_2), (\pi_{2L},k_{2L})$ and $pos(m_{2i},k_{2i}) = \pi_{2i}$ for i=1,L we infer $pos(m_2,k_1) = \pi_1 = pos(m_{2L},k_1)$. Again (2317) may be applied since $k_i \sqsubset c_2, c_{2L}$(i=1,N); $m_2 \in B_{c_2}$, $m_{2L} \in B_{c_{2L}}$,

$pos(m_2,k_i) = pos(m_{2L},k_i)$ and

$(k_N,c_2,m_2,p_2) \to p_2$ (trivial),

$(k_N,c_{2L},m_{2L},p_2) \to p_{2L}$ (***).

Now $pos(p_2,k_1) = pos(p_{2L},k_1)$ follows, further $pos(p_2,k_2) = pos(n_2,k_2)$ by (**) and thus $pos(p_2,k_1) = pos(n_2,k_1) = pos(n_o,k_1)$, using (*). Similarly, $pos(p_{2L},k_1) = pos(n_{2L},k_1)$ holds, from which $pos(n_o,k_1) = pos(n_{2L},k_1)$ is concluded. □

Now consider a triple (π_1,π_2,k) where $k \in B$ and $\pi_1, \pi_2 \in Pos_\varphi(k)$. We will define a map

(2322) $\bar{T}(\pi_1,\pi_2,k) : Pos_\varphi \to Pos_\varphi$.

Let $(\rho_1,l) \in Pos_\varphi$. According to (2315) there exists a 5-chain σ joining (π_1,k) and (ρ_1, l), say

$\sigma \equiv (\pi_1,k) = (\alpha_1,l_1)\ I_1\ \dots\ (\alpha_5,l_5) = (\rho_1,l)$.

Since σ is weakly transportable (2314), there exists a congruent 5-chain τ

$\tau \equiv (\pi_2,k) = (\beta_1,l_1)\ I_1\ \dots\ (\beta_5,l_5) = (\beta_5,l)$.

We will show that the following assignement is well-defined:

(2323) $\bar{T}(\pi_1,\pi_2,k)(\rho_1,l) \underset{def}{=} (\beta_5,l)$

To this end assume another 5-chain

$\sigma' \equiv (\pi_1,k) = (\alpha_1',l_1')\ I_1'\ \dots\ (\alpha_5',l_5') = (\rho_1,l)$

and its transported chain

$\tau' \equiv (\pi_2,k) = (\beta_1',l_1)\ I_1'\ \dots\ (\beta_5',l_5') = (\beta_5',l)$.

We combine the inverse of σ and σ' obtaining the cyclic 9-chain.

$\bar{\sigma} \equiv (\rho_1,l) = (\alpha_5,l_5)\ I_4^{-1}(\alpha_4,l_4)\ \dots\ (\alpha_1,l_1) =$

$(\pi_1,k_1) = (\alpha_1',l_1')\ I_1'\ \dots\ (\alpha_5',l_5') = (\rho_1,l)$.

The corresponding congruent 9-chain $\bar{\tau}$ is cyclic by (2320). Hence

$\beta_5 = \beta_5'$. □

Mappings of the form (2322), (2323) will be called <u>transports</u>, the set of all transports will be denoted by $\bar{T}$.

(2324) Proposition:

(i) Let $\tau \in \bar{T}$. $\tau(\rho_1,l) = (\rho_2,l) \Leftrightarrow \tau = \bar{T}(\tau_1,\tau_2,l)$.

(ii) $\bar{T}(\rho_2,\rho_3,l) \circ \bar{T}(\rho_1,\rho_2,l) = \bar{T}(\rho_1,\rho_3,l)$.

(iii) $\tau_1, \tau_2 \in \bar{T} \Rightarrow \tau_1 \circ \tau_2 \in \bar{T}$,

(iv) $\bar{T}(\pi_1,\pi_1,k) = 1_{Pos}$,

(v) $\bar{T}(\pi_1,\pi_2,k)^{-1} = \bar{T}(\pi_2,\pi_1,k)$.

(vi) $\tau \in T$, $(\pi_1,k) \sqsubset (\rho_1,l) \Rightarrow \tau(\pi_1,k) \sqsubset \tau(\rho_1,l)$.

(2325) Corollary: $\bar{T}$ is a subgroup of Bij(Pos) - the group of all bijections on Pos - and operates freely (see [BGT] III § 4.3 Def. 2) on Pos.

Proof of (2324):

(i) "⇐" follows, since cyclic 5-chains are transportable (2319).

"⇒" Let $\tau = \bar{T}(\pi_1,\pi_2,k)$ and $(\sigma_1,m) \in$ Pos arbitrary. α_1, τ_1, σ_1 can be joined by a cyclic 13-chain σ, such that σ contains 5-chains joining these positions pairwise. According to $\bar{T}(\pi_1,\pi_2,k)(\rho_1,l) = (\rho_2,l)$ and (2320) there exists a congruent cyclic 13-chain τ joining ρ_2, π_2 and some position $\sigma_2 \in$ Pos(m). This proves

$(\sigma_2,m) = \bar{T}(\pi_1,\pi_2,k)(\sigma_1,m) = \bar{T}(\rho_1,\rho_2,l)(\sigma_1,m)$.

(ii) Both mappings yield the same position if applied to some $(\pi_1,k) \in$ Pos.

(iii) Let $\tau_2 = \bar{T}(\rho_1,\rho_2,l)$ and $\tau_1(\rho_2,l) = (\rho_3,l)$. Then, by (i), $\tau_1 = \bar{T}(\rho_2,\rho_3,l)$ and, by (ii), $\tau_1 \circ \tau_2 = \bar{T}(\rho_1,\rho_3,l)$.

(iv) $\bar{T}(\pi_1,\pi_1,k)(\rho_1,l) = (\rho_1,l)$.

(v) Follows from (ii) and (iv).

(vi) Let $\tau = T(\sigma_1,\sigma_2,m)$. Any 5-chain joining (σ_1,m) and (ρ_1,l) also joins (σ_1,m) and (π_1,k). The corresponding congruent chain contains $\tau(\pi_1,k) = (\pi_2,k) \sqsubset (\rho_2,l) = \tau(\rho_1,l)$. □

Next we have to extend transports to spatial regions. Obviously we need the property that transports map equivalent positions (see def. (2219)) onto equivalent positions.

(2326) Lemma: Let (π_i,k), $(\rho_i,l) \in$ Pos (i=1,2) such that $\bar{T}(\pi_1,\pi_2,k)(\rho_1,l) = (\rho_2,l)$ and $\pi_1 \sim \rho_1$. Then $\pi_2 \sim \rho_2$ holds.

(2327) Corollary: $\pi_1 \sim \rho_1 \Rightarrow \pi_2 \sim \rho_2$.

Proof: According to (2219) let $p_2 \in \pi_2$ be given such that the substitution condition holds. Then we have to construct a substitution $(p_2, k, \rho_2, \alpha, r_2, l)$. This will be accomplished in 5 steps.

(STEP1) By axiom B8 (237) there exists a $\bar{p}_2 \in B_{RN}$ such that $p_2 \not\subset \bar{p}_2$. The position of all bodies from $C(p_2)$ remains unchanged. Let $C(\bar{p}_2) = \{c \mid c \sqsubset \hat{p}_2\}$.

(STEP2) Apply $\bar{T}(\pi_2, \pi_1, k)$, i.e. consider $(k, \hat{p}_2, \bar{p}_2, p_1) \to \bar{p}_1$ where $p_1 \in \pi_1$. Thus $pos(p_1, k) = pos(p_1, k) = \pi_1$.

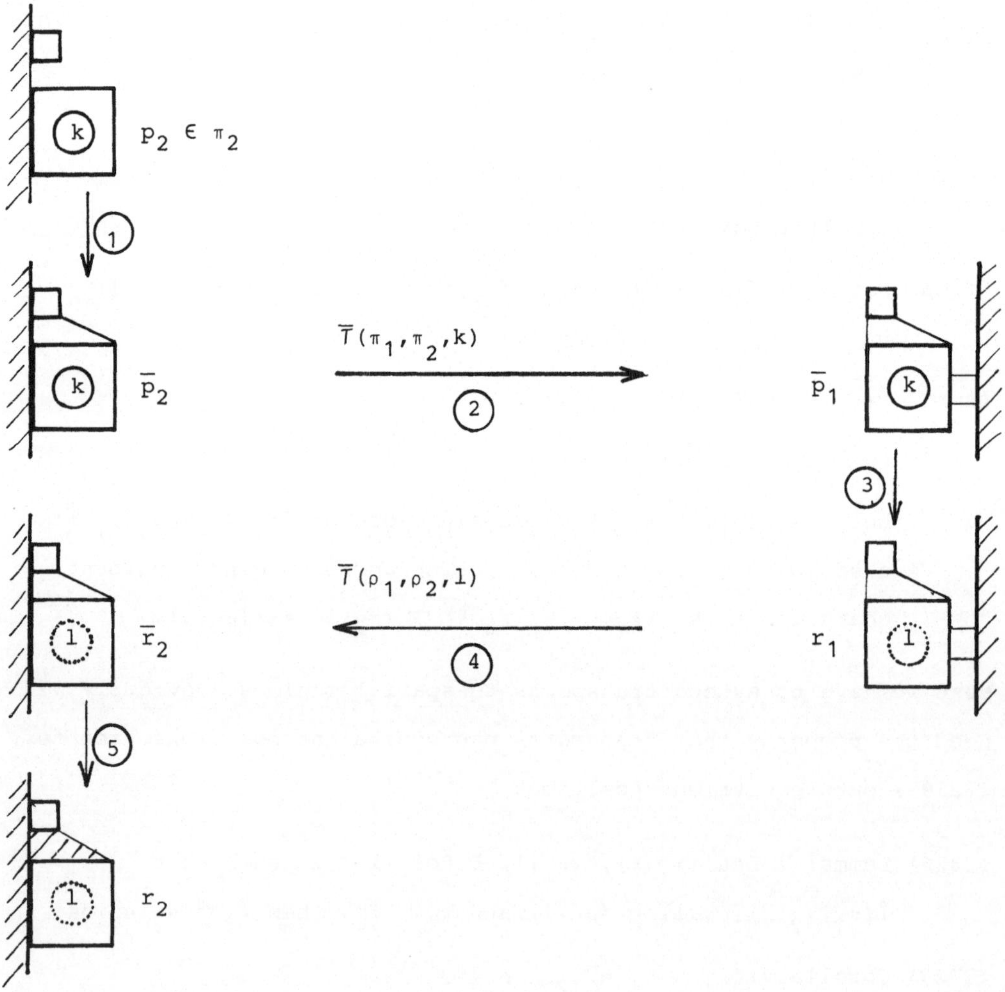

fig.(2328)

(STEP3) We may take the substitution condition w.r. to $\pi_1 \sim \rho_1$ and $\bar{p}_1$ for granted. (Otherwise this could be accomplished by some auxiliary substitution as in section 2.2.). Hence a substitution $(\bar{p}_1, k, \rho_1, \beta, r_1, l)$ exists.

(STEP4) Now consider $(l, \beta\hat{p}_2, r_1, s_2) \to r_2$, where $s_2 \in \rho_2$.

(STEP5) By axiom B6 (236) there exists an $r_2 \in B_R$ such that $C(r_2) = \beta C(p_2,k) \cup \{c | c \sqsubset l\}$ and all bodies of this set have the same position as in $\bar{r}_2$. We set $\alpha \underset{def}{=} \beta | C(p_2,k)$.

It is clear that, by construction, $C(r_2,l) = \alpha C(p_2,k)$ and α, as a restriction of β, enjoys the properties (2214)(ii) c), d), f), which are required for a substitution. It remains to show (2214)(ii) e). Let $m \in C(p_2,k)$ and $m \in C(p_2,k)$ and $m \,\Delta\, k$. We set

$\mu_2 \underset{def}{=} pos(p_2,m) = pos(\bar{p}_2,m)$ and conclude

$(\mu_1,m) \underset{def}{=} \overline{pos}(\bar{p}_1,m) = \overline{T}(\pi_2,\pi_1,k)(\mu_2,m)$. Further $\beta m = m$ and $pos(\bar{p}_1,m) = pos(r_1,m)$, since β is a substitution map. Finally, $\overline{pos}(r_2,m) = \overline{pos}(\bar{r}_2,m) = \overline{T}(\rho_1,\rho_2,l) \circ \overline{T}(\pi_2,\pi_1,k)(\mu_2,m) = (\mu_2,m)$, hence $pos(r_2,m) = pos(p_2,m)$. □

The extension of transports is now a consequence of the following lemma, which can be proved immediately.

(2329) Lemma: Let be $\sim$ an equivalence relation defined on a set P and $f : P \to P$ a map satisfying $\forall\, p,q \in P,\ p \sim q \Rightarrow fp \sim fq$. Then there exists a unique map $\tilde{f} : P/\sim \to P/\sim$ such that the diagram

$$\begin{array}{ccc} P & \xrightarrow{f} & P \\ \downarrow \nu & & \downarrow \nu \\ P/\sim & \xrightarrow{\tilde{f}} & P/\sim \end{array}$$

commutes (ν denoting the canonical surjection).

If $g : P \to P$ is another map, we have $\widetilde{f \circ g} = \tilde{f} \circ \tilde{g}$ and

$\widetilde{1_P} = 1_{P/\sim}$. If f is bijective, so is $\tilde{f}$ and $\widetilde{f^{-1}} = \tilde{f}^{-1}$.

We will write

$\overline{T}(\pi_1,\pi_2,k) = \widetilde{T(\pi_1,\pi_2,k)} : R \to R$ and denote by T the set of <u>transport mappings</u> obtained in this way.

(2330) Theorem: T is a subgroup of $\mathrm{Aut}(R,\prec)$.

Proof: By (2329), T is a subgroup of $\mathrm{Bij}(R)$. If remains to show, that $A \prec B$ implies $\tau A \prec \tau B$ for all $\tau \in T$ and $A, B \in R$. But this follows immediately from (2324)(vi). □

Finally we are to lift transport mappings to the lattice of regions $(R,<)$.

(2331) Definition:

(i) Let $\tau \in T$ and $a = \{A_1 \ldots A_n\} \in R$, $\hat{\tau}a \underset{\text{def}}{=} \{\tau A_1 \ldots \tau A_n\}$. Since $\tau \in \mathrm{Aut}(R,\prec)$, $A_i \vartriangle A_j$ implies $\tau A_i \vartriangle \tau A_j$. Hence $\hat{\tau}a \in R$ and $\hat{\tau} : R \to R$.

(ii) $T \underset{\text{def}}{=} \{\hat{\tau} \mid \tau \in T\}$. The elements of T are also called transport mappings.

(2332) Theorem: T is a subgroup of $\mathrm{Aut}(R,<)$.

Proof: Let be $a, b \in R$, $a < b$ and $\tau \in T$. $a < b$ means:

$\forall A \in a \; \exists B \in b$ such that $A \prec B$.

It follows, that

$\forall \tau A \in \hat{\tau}a \; \exists \tau B \in \hat{\tau}b$ such that $\tau A \prec \tau B$. Hence $\hat{\tau}a < \hat{\tau}b$. The remainder of the claim follows from $\widehat{(\sigma \circ \tau)} = \hat{\sigma} \circ \hat{\tau}$ and $\hat{1}_R = 1_R$. □

3. REGIONS AND TRANSPORT MAPPINGS

We will formulate a theory PT_2 whose species of structure is given by a principal base set R, the structural term which is characterized by

$(<,T) \in P(R\times R) \times PP(R\times R)$

and axioms R1 to R6. (Later in section 5 the axioms R7 and R8 will be added.)

The theory PT_1 in the preceding section is to be viewed as a "pre-theory" of PT_2, which establishes the physical interpretation of the elements:

$a \in R$ as regions,

$\tau \in T$ as transport mappings and of

$a < b$ as an inclusion of regions.

More precisely, we state, using the terminology of [LUD 3], that the theory PT_2' results from PT_1 by restriction (Einschränkung) and embedding (Einbettung), where PT_2' is PT_2 without axioms R3 to R8. In other words: R1 and R2 are already proven as theorems within PT_1.

We prefere to formulate PT_2 as an independent theory thus admitting the possibility of other pre-theories.

It should be emphasized, that the structures "region", "transport mapping" and "inclusion" are highly idealized constructs which have been developed via a process of abstraction from simple concepts, which refer to in the concrete observable entities, as "body", "transport" and "part of".

3.1 4 AXIOMS

(311) Axiom R1: $(R,<)$ is a weakly distributive lattice with least element O and $R \neq \{O\}$
(see def. (2244) and compare with prop. (2246)).

This means that R is partially ordered by the relation <, and for any two regions v, w ∈ R there exist supremum $v \vee w$ and infimum $v \wedge w$, which may well be the "empty region" O.

The supremum and infimum of finitely many regions will be denoted by $\bigvee_{i=1}^{n} a_i$ and $\bigwedge_{i=1}^{n} a_i$ resp.; R_o is the subset of "non-empty" regions; Aut R the group of all order preserving automorphisms of R.

(312) Axiom R2: T is a subgroup of Aut R (compare prop. (2332)).

(313) Axiom R3:

$\forall\, v \in R_o\; \exists\, w \in R_o,\; w < v$ such that

$\forall\, \tau \in T,\; w \wedge \tau w \neq O \Rightarrow \tau w < v.$

"Each region contains a subregion and each of its displacements which still intersects the subregion".

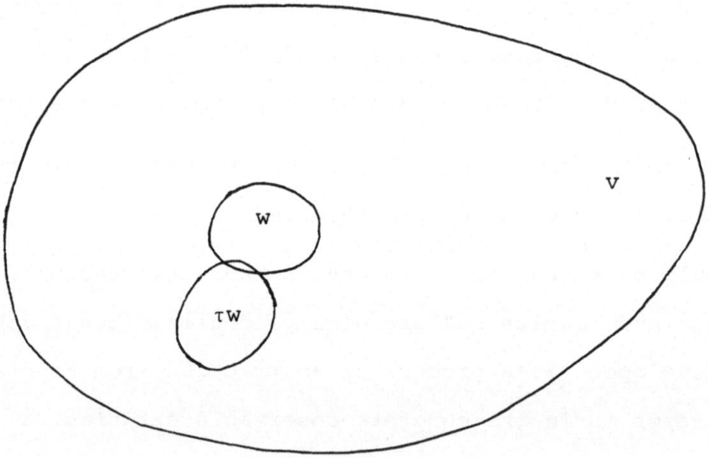

fig. (314)

In this case we call w a kernel of v (re T). Axiom R3 does not mean, that the underlying space is infinitely divisible. There exist models in which R has atoms w, thus fulfilling

$\forall\, \tau \in T$, either $w \wedge \tau w = O$ or $w < \tau w$ holds, (see (318),(3110),(336)).

As a consequence of axiom R3, T can be endowed with the structure of a topological group.

(315) Definition:

$$T(a) \underset{def}{=} \{\tau \in T \mid \tau a \wedge a \neq 0\},\ a \in R$$

$$\mathcal{U} \underset{def}{=} \{T(a) \mid a \in R_o\}.$$

T(a) should not be confused with Ta = $\{\tau a \mid \tau \in T\}$.

(316) Proposition:

$\{\varphi U \mid \varphi \in T, U \in \mathcal{U}\}$ is an open sub-base of a topology t on T.

(T,t) is a topological group.

Proof: By [BGT] III, § 1.2 the following 4 points need to be shown:

(a) Let FI($\mathcal{U}$) denote the set of finite intersections in $\mathcal{U}$, then $\emptyset \notin$ FI($\mathcal{U}$). This holds since $\forall\ U \in \mathcal{U}$, $Id_R \in U$.

(b) $\forall\ U \in \mathcal{U}\ \exists\ V \in FI(\mathcal{U})$ such that $V^2 \subset U$.

Let U = T(a), a $\in R_o$, b kernel of a, c kernel of b,

V = T(c) $\subset$ T(b).

For any τ, $\sigma \in$ V it follows:

$\sigma\ c \wedge c \neq o$, $\tau\ b \wedge b \neq o$

$\sigma\ c < b$, $\quad \tau\ b < a$

$\tau\ \sigma\ c < \tau\ b,\ a$, $\tau\ \sigma\ a$

$\tau\ \sigma\ a \wedge a \neq o$

$\tau\ \sigma \in T(a) = U$.

Hence: $V^2 \subset U$.

(c) $\forall\ U \in \mathcal{U}\ \exists\ V \in \mathcal{U}$ such that $V^{-1} \subset U$.

Since $\tau\ a \wedge a \neq o \Leftrightarrow \tau^{-1}\ a \wedge a \neq o$, even $U^{-1} = U$ holds.

(d) $\forall\ U \in \mathcal{U}\ \ \forall\ \chi \in T\ \exists\ V \in \mathcal{U}$ such that $\chi\ V\ \chi^{-1} \subset U$.

Let U = T(a) and choose V = $T(\chi^{-1}a)$.

For any $\psi \in V$, $\chi \psi \psi^{-1} a \wedge a \neq o \Leftrightarrow \psi(\chi^{-1} a) \wedge \chi^{-1} a \neq o$, therefore: $\chi V \chi^{-1} = U$.

Finally, $\mathcal{U}$ consists of open sets if the following is true:

"$\forall U \in \mathcal{U} \quad \forall \chi \in U \; \exists V \in \mathcal{U}$ such that $\chi V \subset U$".

Let $U = T(a)$ and $\chi \in U$, that is $c \underset{def}{=} \chi a \wedge a \neq o$. Choose $V \underset{def}{=} T(b)$, where $b \underset{def}{=} \chi^{-1} c = a \wedge \chi^{-1} a$. Clearly $b, c < a$.

For any $\varphi \in \chi V$ it follows by definition of V

$d \underset{def}{=} \varphi b \wedge \chi b \neq o$

$d < \chi b = c < a$ and $d < \varphi b < \varphi a$

$d < a \wedge \varphi a \neq o$

$\varphi \in T(a) = U$.

Hence: $\chi V \subset U$. □

(317) Axiom R4:

$\forall a \in R \; \forall v \in R_o, \; \exists n \in \mathbb{N} \; \exists \tau_1, \ldots, \tau_n \in T$

such that $a < \bigvee_{i=1}^{n} \tau_i v$.

"Each region can be covered by finitely many regions of arbitrarily small order".

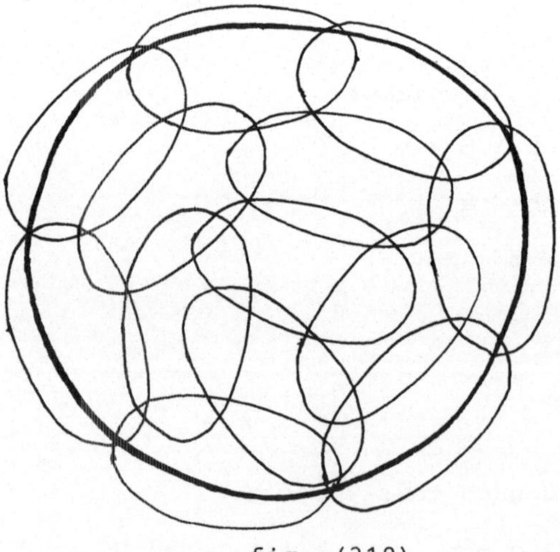

fig. (318)

Examples of such tripels $(R,<,T)$, fulfilling the axioms R1 to R4, can be taken from the various geometries in section 4.2, where $(R,<)$ may be chosen as the lattice of all relatively compact, open subsets. Thus, we may well confine ourselves to a simple

(319) Example: $(R,<)$ is taken as the lattice of finite joins of real, bounded, closed (resp. open, resp. open-closed) intervals, T as the group of lattice automorphisms induced by translations of the real line.

The independence of the essential axioms is made clear by the following counter-examples. The postulate, which does <u>not</u> hold, is given in brackets.

(3110) Counter-example (weak distributivity):
Let $(R,<)$ be the lattice of all linear subspaces of $\mathbb{R}^2$, T the group induced by rotations.

(3111) Counter-example (axiom R3):
$(R,<)$ is the ordered set of the nonnegative reals, T consists of all dilatations $x \mapsto \alpha \cdot x$, $\alpha > o$.

(3112) Counter-example (axiom R4):
$(R,<)$ is the lattice of continuous, nonnegative, real functions with

$$f < g \underset{\text{def}}{\Leftrightarrow} \forall\, x \in \mathbb{R},\ f(x) \leq g(x).$$

T is the group of automorphisms τ_α, induced by translations:

$$\tau_\alpha(f)(x) \underset{\text{def}}{=} f(x-\alpha),\ \alpha \in \mathbb{R}.$$

(3113) Counter-example (axiom R3):
$(R,<)$ is the lattice of bounded, open subsets of $\mathbb{R}^{n+1}$, T is

induced by the Poincaré-group of $\mathbb{R}^{n+1}$.

The counter-examples (3111) and (3112) show, that axiom R3 may be loosened without menacing the continuity of multiplication in T. Instead of "T(w) w < v" it would suffice to postulate " $\bigcap_{i=1...n} T(w_i)\, w_j < v$" which leads to the concept of a "multi-kernel" $(w_1,...,w_n)$. A theory for (R,<,T), adopting this weakened version of R3 to allow for space-time-geometry , has recently been worked out by D. Mayr [MAY].

3.2 POINTS

The intuitive idea of points as "very small regions" can be made precise through our construction of points. It makes possible a representation of regions as subsets of points. We have to fulfil the requirement, that, if one starts with a space of points and a lattice of "resonable" subsets, the construction should permit one to recapture the given space.
This condition excludes a purely lattice-theoretical construction like [BGT] I §6, Ex. 17-18, as is shown by the counter-example (329) below. It seems necessary to use the surrogate of a uniform structure present in the group T in order to imitate the process of completion of a uniform space.

(321) Definition:

A subset $F \subset R$ satisfying

(i) $O \notin F$, and

(ii) $\forall\, a, b \in F\ \exists\, c \in F$ such that $c < a \wedge b$, is said to be a <u>prefilter basis</u>. If additionally,

(iii) $\forall\, a \in F\ \forall\, b \in R,\ b > a \Rightarrow b \in F$ holds, F is called a <u>prefilter</u>.

A prefilter F satisfying

(iv) $\forall\, a \in R_o\ \exists\, \tau \in T$ such that $\tau a \in F$ is called a <u>Cauchy prefilter</u>.

In the event the prefilter

(v) $F \subset G$ (set-th. inclusion)

we will say, that G is <u>finer</u> than F and F is <u>coarser</u> than G. A prefilter U satisfying

(vi) $\forall\, F$, F prefilter, such that $U \subset F \Rightarrow U = F$, is called <u>maximal</u>, and a prefilter P for which

(vii) $\forall\, a, b \in R$, $a \vee b \in P \Rightarrow a \in P$ or $b \in P$

holds, is called <u>prime prefilter</u>.

As in the case with filters, whose elements are subsets, one can easily prove the following

(322) Proposition: Let $\mathcal{M} \neq \emptyset$ be a family of prefilters.

(i) $\underline{\mathcal{M}} \underset{def}{=} \bigcap_{F \in \mathcal{M}} F$ is a prefilter,

(ii) $\overline{\mathcal{M}} \underset{def}{=} \bigcup_{F \in \mathcal{M}} F$ is a prefilter, if $\mathcal{M}$ is linearly ordered.

(iii) For any prefilter F there exists a maximal prefilter U, which is finer than F.

(323) Lemma: Let U be a maximal prefilter.

(i) If $a \in U$, $b \in R_o$, $b < a$, then either $b \in U$ or $\exists\, c \in U$ such that $b \wedge c = o$.

(ii) U is a prime prefilter. (If R is distributive, see [BGT] I §6, Ex. 18b.)

Proof:

(i) Assume $\forall\ c \in U,\ b \wedge c \neq o$. In this case $U \cup \{b\}$ is a prefilter base and generates a prefilter which is finer than U. But U is maximal, hence $U \cup \{b\} = U$, i. e. $b \in U$.

(ii) Assume $a_1 \vee a_2 \in U$, but $a_1 \notin U$ and $a_2 \notin U$. According to (i) there exist $b_1, b_2 \in U$ such that $a_1 \wedge b_1 = 0 = a_2 \wedge b_2$. Thus neither a_1 nor a_2 meet $b_1 \wedge b_2$, and, by weak distributivity (311), $0 = (a_1 \vee a_2) \wedge (b_1 \wedge b_2)$, which is impossible because U is a prefilter. □

(324) Corollary:

$$\bigvee_{i=1}^{n} a_i \in U \Rightarrow \exists\ i \in \{1,\dots,n\} \text{ such that } a_i \in U.$$

(325) Proposition: Each maximal prefilter U is a Cauchy prefilter.

Proof: Let be $a \in U$, $b \in R_o$. According to (317) a can be covered by finitely many $\tau_i\, b$, $\tau_i \in T$:

$a < \bigvee_{i=1}^{n} \tau_i\, b \in U$. By (324) some $\tau_i\, b \in U$, which is just the Cauchy property. □

(326) Proposition: The family $\check{C}$ of all Cauchy prefilters is inductively ordered downwards (i.e., any linearly ordered subfamily M has an infimum).

Proof: Take $G \underset{\text{def}}{=} \bigcap_{F \in M} F$. We have to show $G \in \check{C}$. Clearly G is a prefilter (322)(i). Let be U any maximal prefilter $U \supset \bar{M}$, $a \in R_o$ and b a kernel of a (see def. (314)). Since U is a Cauchy prefilter (325), we have $\sigma\, b \in U$ with some $\sigma \in T$. Likewise there exists some $\tau_F \in T$ such that $\tau_F\, b \in F$ for all $F \in M$. We conclude:

$U \supset \bar{M} \supset F$

$\tau_F\ b \in U,\ \sigma\ b \in U$

$\tau_F\ b \wedge \sigma\ b \neq 0$

$\sigma^{-1}\ \tau_F\ b \wedge b \neq 0$

$\sigma^{-1}\ \tau_F\ b < a$

$\tau_F\ b < \sigma\ a$

$\sigma\ a \in F.$

This holds for all $F \in M$, hence: $\sigma\ a \in G$. □

From (326) it follows by Zorn's lemma that for each Cauchy prefilter F there exists a minimal Cauchy prefilter G, which is coarser than F.

(327) Definition: The elements of the set of all minimal Cauchy prefilters, P, will be called <u>points</u>.

(328) Proposition: For each Cauchy prefilter F there exists a unique minimal Cauchy prefilter $x \in P$, which is coarser than F. In this case we will write: $F \to x$ or $\lim F = x$.

Proof: Let be $x, y \in P$ two points, which are coarser than $F : x, y \subset F$. We have to show $x = y$.
Take any $a \in R_o$ and b as a kernel of a.
Then the following holds:

$\exists\ \sigma, \tau \in T$ such that $\sigma\ b \in x,\ \tau\ b \in y$

$\sigma\ b,\ \tau\ b \in F$

$\sigma\ b \wedge \tau\ b \neq 0$

$\tau^{-1}\ \sigma\ b \wedge b \neq 0$

$\tau^{-1}\ \sigma\ b,\ b < a$

$\tau\ b,\ \sigma\ b < \tau\ a$

$\tau\ a \in x, y.$

Now we choose a $\tau \in T$ for each $a \in R_o$ and consider the family of all τ a. Finitely many regions τ_i a_i have a nonempty intersection, because they are contained in the prefilter x. Hence the family of all τ a generates a prefilter H, which is Cauchy by its very construction. The above assertion τ a $\in$ x, y shows that H is coarser than x and y. But these were assumed to be minimal, hence:
x = H = y. □

It would be convenient at this point to make some comments not used in sequel.
Alternative candidates for points are the prime prefilters or the subset of maximal prefilters. The following (counter-) example however shows that this construction would be unnatural in a physical setting.

(329) Counter-example: Let be (R,<) the lattice of finite unions of real intervals of the form $\bigcup_{\nu=0\ldots n} I_{2\nu}$ with $I_{2\nu} =]a_{2\nu},a_{2\nu+1}[$ or $[a_{2\nu},a_{2\nu+1}[$ or $]a_{2\nu},a_{2\nu+1}]$ or $[a_{2\nu},a_{2\nu+1}]$, satisfying $a_\lambda < a_{\lambda+1}$, $\lambda = 0,\ldots 2n+1$.

Since $]a,b[\cup]b,c[$ does not occur in R, we have $]a,b[\vee]b,c[=]a,c[$ and $\{]x-\varepsilon,x+\varepsilon[\,|\,\varepsilon>0\}$ does not form the base of a prime prefilter.

The only prime prefilter bases are of the form $\{]x,x+\varepsilon[\,|\,\varepsilon>0\}$ and $\{]x-\varepsilon,x[\,|\,\varepsilon>0\}$, where $x \in \mathbb{R}$. (These even generate maximal prefilters.) Hence the space of prime (resp. maximal) prefilters Ω is isomorphic to $\mathbb{R} \times \{1,2\}$.
By the way, this is an example of a lattice, which is weakly distributive, but not distributive. The latter follows from

$]a,b] \wedge (]a,b[\vee]b,c[) =]a,b]$ and

$(]a,b]\wedge]a,b[) \vee (]a,b]\wedge]b,c[) =]a,b[$.

If we restrict the example to open intervals $I_{2\nu} =]a_{2\nu},a_{2\nu+1}[$, $(R,<)$ becomes distributive without altering Ω. If we replace $a_\lambda < a_{\lambda+1}$ by $a_\lambda \leq a_{\lambda+1}$, there are additional prime (but not maximal) prefilters, generated by $\{]x-\varepsilon,x+\varepsilon[\,|\,\varepsilon>0\}$. Now, $\Omega \cong \mathbb{R} \times \{1,2,3\}$. Thus, it becomes clear, how sensitive this concept of a "point" depends on the arrangement of the boundary of regions. For this reason this concept must be ruled out for a physical geometry.

3.3 REGIONS AS POINT SETS

(331) Definition:

(i) For each $a \in R$ we define

$$\tilde{a} \underset{def}{=} \{x\in P\,|\,a\in x\}.$$

Hence $x \in \tilde{a} \Leftrightarrow a \in x$.

$$\tilde{R} \underset{def}{=} \{\tilde{a}\,|\,a\in R\}.$$

(ii) Let be $a \in R_o$.

$$N_a \underset{def}{=} \{(x,y)\in P\times P\,|\,\exists\tau\in T \text{ such that } x,y\in\tau a\}.$$

$$\mathcal{N} \underset{def}{=} \{N_a\,|\,a\in R_o\}.$$

(332) Proposition:

$\forall\, a \in R_o\ \exists\, x \in P$ such that $x \in \tilde{a}$.

"Each nonempty region contains at least one point".

Proof: Let be $b \in R_o$ a kernel of a, F the prefilter generated by $\{b\}$, U a maximal prefilter with $U \supset F$ and $x = \lim U$. It follows:

x Cauchy prefilter

$\exists\ \tau \in T$ such that $\tau\, b \in x$

$\tau\, b,\ b \in U$

$\tau b \wedge b \neq 0$

$\tau b < a$

$a \in x$

$x \in \tilde{a}$. □

(333) Proposition: N is a fundamental system of entourages of a uniformity on P (see [BGT] II § 1.1).

Proof: We have to prove 4 items:

(B_I) in [BGT] I § 6.3: "$\forall\, a, b \in R_o \;\; \exists\, d \in R_o$ such that $N_d \subset N_a \cap N_b$". Choose $x \in \tilde{a}$ (332) and $\tau \in T$ such that $\tau b \in x$. Set

$d \underset{def}{=} a \wedge \tau b \neq 0$ and take any $(y,z) \in N_d$. It follows:

$\exists\, \varphi \in T$ such that $\varphi d \in y, z$

$\varphi a \wedge \varphi \tau b \in y, z$

$\varphi a \in y, z$ and $\varphi \tau b \in y, z$

$(y,z) \in N_a$ and $(y,z) \in N_b$

$(y,z) \in N_a \cap N_b$.

(U'_I) in (BGT) II § 1.1:

"$\forall\, a \in R_o \;\; \forall\, x \in P, \;\; (x,x) \in N_a$".

Since x is Cauchy, there exists a $\tau \in T$ such that $\tau a \in x$.

(U'_{II}) : "$\forall\, N_a \in N \;\exists\, N_b \in N$ such that $N_b \subset N_a^{-1}$".

This holds trivially since $N_a = N_a^{-1}$.

(U'_{III}) : "$\forall\, a \in R_o \;\exists\, b \in R_o$ such that $N_b \circ N_b \subset N_a$".

Choose b as a kernel of a and any (x,y), $(y,z) \in N_b$. It follows:

$\exists\, \tau, \sigma \in T$ such that $\tau b \in x, y$ and $\sigma b \in y, z$

$\tau b \wedge \sigma b \neq 0$, since y is a prefilter

$\sigma^{-1} \tau b \wedge b \neq 0 \Rightarrow \sigma^{-1} \tau b < a$

$\tau b < \sigma a$, $\sigma b < \sigma a$

$\sigma a \in x, z$

$(x,z) \in N_a$. □

In sequel we will always assume P to be endowed with the uniform structure given by $\mathcal{N}$.

(334) Proposition: P is a uniform Hausdorff space (hence regular, see [BGT] II, § 1.2, Prop. 3).

Proof: Let be $(x,y) \in \bigcap_{a \in R_o} N_a$, which means

$\forall a \in R_o \ \exists \sigma \in T$ such that $\sigma a \in x, y$.

The family of all σa has the finite intersection property, because all σa are elements of the prefilter x. Hence they generate a prefilter F, which is Cauchy by construction and coarser than x and y. By minimality of x, y we infer $x = F = y$. □

(335) Definition:

(i) Let be $a \in R_o$.

$$\bar{\bar{N}}_a \underset{def}{=} \{(M,N) \in \mathcal{P}P \times \mathcal{P}P \mid M \subset N_a(N), N \subset N_a(M)\},$$

$$\bar{\bar{\mathcal{N}}} \underset{def}{=} \{\bar{\bar{N}}_a \mid a \in R_o\}.$$

(ii) Let be $M, N \subset P$.

$$M \sim N \underset{def}{\Leftrightarrow} \forall a \in R_o, (M,N) \in \bar{\bar{N}}_a.$$

$\bar{\bar{\mathcal{N}}}$ is a fundamental system of entourages of a uniformity on $\mathcal{P}$ P([BGT] II § 1, Ex. 5), which is in general non-Hausdorff. In this case the equivalence relation $\sim$ is not the identity. Since $\bar{M} = \bigcap_{a \in R_o} N_a(M)$ we have: $M \sim N \Leftrightarrow \bar{M} = \bar{N}$.

The assignement $a \mapsto \tilde{a}$ is a lattice representation in the following weak sense.

(336) Proposition: Let be $a, b \in R$.

(i) $a < b \Rightarrow \tilde{a} \subset \tilde{b}$,

(ii) $\tilde{a} = \emptyset \Leftrightarrow a = 0$,

(iii) $\widetilde{a \wedge b} = \tilde{a} \cap \tilde{b}$,

(iv) $\widetilde{a \vee b} \supset \tilde{a} \cup \tilde{b}$,

(v) $\widetilde{a \vee b} \sim \tilde{a} \cup \tilde{b}$, see (335)(ii),

Define for $\mathcal{A} \subset \mathcal{P}\, P$: $\underset{\sim}{\mathcal{A}} \underset{\text{def}}{=} \{a \in R \mid \tilde{a} \in \mathcal{A}\}$.

(vi) $\mathcal{C}$ filter on P $\Rightarrow$ $\underset{\sim}{\mathcal{C}}$ prefilter,

(vii) $\mathcal{C}$ Cauchy filter on P $\Rightarrow$ $\underset{\sim}{\mathcal{C}}$ Cauchy prefilter,

(viii) F prefilter $\Rightarrow$ $\tilde{F}$ filter base on P,

(ix) F Cauchy prefilter $\Rightarrow$ $\tilde{F}$ Cauchy filter base on P,

(x) $\mathcal{C}$ minimal Cauchy filter on P $\Rightarrow$

$\underset{\sim}{\mathcal{C}}$ minimal Cauchy prefilter (i.e. point).

Proof:

(i) $x \in \tilde{a}$

$a \in x$

$b \in x$, since x is a prefilter

$x \in \tilde{b}$.

(ii) $a \neq 0 \Rightarrow \tilde{a} \neq \emptyset$ by (332).

$x \in \tilde{0} \Rightarrow 0 \in x$ in contradiction to "x is a prefilter".

(iii) $a \wedge b < a, b$

$\widetilde{a \wedge b} \subset \tilde{a}, \tilde{b}$ by (i)

$\widetilde{a \wedge b} \subset \tilde{a} \cap \tilde{b}$.

$x \in \tilde{a} \cap \tilde{b}$

$a, b \in x$

$a \wedge b \in x$, since x is a prefilter

$x \in \widetilde{a \wedge b}$.

(iv) $a \vee b > a, b$

$\widetilde{a \vee b} \supset \tilde{a}, \tilde{b}$ by (i)

$\widetilde{a \vee b} \supset \tilde{a} \cup \tilde{b}$.

(v) In view of (iv) it suffices to prove $\widetilde{a \vee b} \subset \bar{\bar{N}}_v(\tilde{a} \cup \tilde{b})$ for arbitrary $v \in R_o$.

Let $x \in \widetilde{a \vee b}$ and u be a maximal prefilter with $u \supset x$. From $a \vee b \in x$, u and "u prime prefilter" (323)(ii) we infer: $a \in u$ or $b \in u$. Since $\sigma v \in x, u$ for some $\sigma \in T$, we conclude $\sigma v \wedge a \neq 0$ or $\sigma v \wedge b \neq 0$. By (332) there exists a $y \in \widetilde{\sigma v \wedge a} \cup \widetilde{\sigma v \wedge b} \subset \tilde{a} \cup \tilde{b}, N_v(x)$, hence $x \in \bar{\bar{N}}_v(\tilde{a} \cup \tilde{b})$.

(vi) Follows from (i), (ii) and (iii).

(vii) The Cauchy property of $\mathcal{C}$ means:

$\forall a \in R_o \; \exists C \in \mathcal{C}$ such that $C \times C \subset N_a$, that is:

$\forall a \in R_o \; \exists C \in \mathcal{C} \; \forall x, y \in C \; \exists \tau \in T$ such that $x, y \in \tau a$.

Now consider any $b \in R_o$. We have to show: $\exists \sigma \in T$ such that $\sigma b \in \tilde{\mathcal{C}}$.

To this end, consider a kernel $a \in R_o$ of b and a subset $C \in \mathcal{C}$ with the property stated above. Choose a point $x \in C$ and $\sigma \in T$ such that $\sigma a \in x$, that is $x \in \sigma a$. For any $y \in C$ there exists some $\tau \in T$ such that $x, y \in \tau a$ (see above). Now we conclude:

$x \in \widetilde{\sigma a}, \widetilde{\tau a}$

$\widetilde{\sigma a} \cap \widetilde{\tau a} \neq \emptyset$

$\sigma a \wedge \tau a \neq 0$

$a \wedge \sigma^{-1} \tau a \neq 0$

$\sigma^{-1}\ \tau\ a < b$

$\tau\ a < \sigma\ b$

$\widetilde{\tau a} \subset \widetilde{\sigma b}$

$y \in \widetilde{\sigma b}$.

Since $y \in C$ was chosen arbitrarily, we have shown:

$C \subset \widetilde{\sigma b}$ and, C being a filter, $\widetilde{\sigma b} \in C$. Equivalently, $\sigma b \in \underset{\sim}{C}$.

(viii) Follows from (i), (ii), (iii).

(ix) Let be $a \in R_o$. There exists a $\sigma \in T$ such that $\sigma\ a \in F$. If we set $C \underset{def}{=} \widetilde{\sigma a}$, the required property $C \times C \subset N_a$ follows.

(x) By (vii) $\underset{\sim}{C}$ is a Cauchy prefilter. Assume $F \subset \underset{\sim}{C}$ is a coarser Cauchy prefilter. According to (viii) $\tilde{F} \subset C$ is a Cauchy filter base on P and hence generates a Cauchy filter G, which is coarser than C. By minimality of C we conclude $G = C$, whence $\underset{\sim}{G} = \underset{\sim}{C}$. It remains to show $\underset{\sim}{G} = F$.

Let be $a \in \underset{\sim}{G}$. It follows:

$\tilde{a} \in G$

$\exists\ \tilde{b} \in \tilde{F}$ such that $\tilde{b} \subset \tilde{a}$

$\exists\ b \in F$ such that $b < a$

$a \in F$.

Let be $a \in F$. From $\tilde{a} \in \tilde{F} \subset G$ we infer $a \in \underset{\sim}{G}$. □

It is clear, that the representation $a \mapsto \tilde{a}$ need not be faithful.

(337) Proposition: P is complete.

Proof: Let C be a Cauchy filter on P.
According to (336)(vii) $\underset{\sim}{C}$ is Cauchy prefilter. Put $x = \lim \underset{\sim}{C}$, hence $\tilde{x} \subset C$. In order to prove $\lim C = x$, it suffices to show that the Cauchy filter base $\tilde{x}$ (336)(ix) is finer than the neighbourhood filter of x. Consider a neighbourhood $N_a(x)$ and find a $\tau \in T$ such

that $\tau a \in x$, $\widetilde{\tau a} \in \tilde{x}$. Since $y \in \widetilde{\tau a} \Rightarrow y \in N_a(x)$, $N_a(x) \supset \widetilde{\tau a}$ and thus $N_a(x)$ is contained in the filter generated by $\tilde{x}$. □

(338) Proposition: Let be $x \in P$.

$\tilde{\tilde{x}} = \{\tilde{a} | a \in x\}$ is an open base of the neighbourhood filter $\mathcal{U}(x)$.

Proof: $\tilde{x}$ generates a filter, which is finer than $\mathcal{U}(x)$, as shown in the proof of (337). Consequently x is finer than $\underset{\sim}{\mathcal{U}}(x)$. Since $\underset{\sim}{\mathcal{U}}(x)$ is Cauchy (336)(vii) and x minimal, we have $x = \underset{\sim}{\mathcal{U}}(x)$ and therefore $\tilde{x} \subset \mathcal{U}(x)$.

This proves $\tilde{x}$ to be a base of $\mathcal{U}(x)$. In order to show "$\tilde{a}$ open" consider any $y \in \tilde{a}$. We infer as above:

$\tilde{a} \in \tilde{y} \subset \mathcal{U}(y)$. Thus $\tilde{a}$ is a neighbourhood of all points $y \in \tilde{a}$ and hence open. □

(339) Proposition: P is locally compact.

Proof: Let be $x \in P$ and $\tilde{a} \in \tilde{x}$ a neighbourhood (338). For any $b \in R_o$ we have a covering

$a < \bigvee_{i=1...n} \tau_i b$, $\tau_i \in T$, by axiom R4.

By (336)(i), (iv) and (v):

$$\tilde{a} \subset \widetilde{\bigvee_{i=1...n} \tau_i b} \sim \bigcup_{i=1...n} \widetilde{\tau_i b}, \text{ hence}$$

$$\tilde{a} \subset \overline{\bigcup_{i=1...n} \widetilde{\tau_i b}} = \bigcup_{i=1...n} \overline{\widetilde{\tau_i b}}.$$ The sets $\tau_i b$ are N_b^2-small.

According to [BGT] II § 4.2 this covering property is equivalent to "$\tilde{a}$ precompact". By completeness of P, $\bar{\tilde{a}}$ will be a compact neighbourhood of x. □

(33310) Definition: $\bar{R} \underset{\text{def}}{=} \{A \subset P | A \text{ is open and relatively compact}\}$.

(3311) Corollary: $\tilde{R} \subset \bar{R}$.

(3312) Proposition: $\tilde{R}$ is a dense subset of $\bar{R}$ w.r. to the topology induced by $\bar{\bar{N}}$ (see (335)).

Proof:

1. Let be $a \in \bar{R}$ and $a \in R_o$. $\bar{A}$ is compact and can be covered by finitely many $\tau_\nu a$, $\nu=1\ldots n$. We may assume $\tau_\nu a \cap A \neq \emptyset$ for all ν. It follows by (336)(v):

$$A \subset \bigcup_{\nu=1\ldots n} \widetilde{\tau_\nu a} \underset{def}{=} B \sim B' \underset{def}{=} \widetilde{\bigvee_{\nu=1\ldots n} \tau_\nu a} \in \tilde{R}.$$

2. It suffices to prove $B \in \bar{\bar{N}}_a(A)$.

2.1. $A \subset N_a(B)$ holds since $A \subset B$.

2.2. $B \subset N_a(A)$. Let be $x \in B$, that is $x \in \widetilde{\tau_\nu a}$ for some $\nu=1\ldots n$. It follows:

$\widetilde{\tau_\nu a} \cap A \neq \emptyset$ (see above)

$\exists\, y \in A$ such that $x, y \in \widetilde{\tau_\nu a}$

$x \in N_a(A)$. □

An equivalent approach to the representation of R by point sets would consist in defining a (in general non-Hausdorff) uniformity on R by entourages $N_R(v)$, $v \in R_o$, of the form:

(3313)

$$(a,b) \in N_1(v) \underset{def}{\Leftrightarrow} \forall\, w \in R \text{ such that } w \wedge a \neq 0 \;\; \exists\, \sigma \in T \text{ such that } w \wedge \sigma v \neq 0 \text{ and } \sigma v \wedge b \neq 0,$$

$$(a,b) \in N_R(v) \underset{def}{\Leftrightarrow} (a,b) \in N_1(v) \text{ and } (b,a) \in N_1(v).$$

The points could be defined as atoms in the Hausdorff completion $\hat{R}$ of R. $\hat{R}$ is isomorphic to the lattice of compact subsets of P. The verification of this approach is left to the reader.

3.4 CONGRUENT MAPPINGS

In an obvious manner transport mappings τ induce mappings of points $\tilde{\tau}: P \to P$, which will be named "congruent mappings". We shall compile some properties of congruent mappings and of the assignement $\tau \mapsto \tilde{\tau}$.

(341) Definition:

(i) A bijection $f : P \to P$ will be called an isometry (of P) iff

$$\forall a \in R_o \ \forall x, y \in P : (x,y) \in N_a \Leftrightarrow (fx,fy) \in N_a.$$

(ii) Let be $x \in P$, $\tau \in T$.

$$\tilde{\tau} x \underset{\text{def}}{=} \{\tau a \mid a \in x\}.$$

(342) Proposition: Let be $\tau \in T$, $x \in P$.

Then $\tilde{\tau}$ is a mapping $\tilde{\tau} : P \to P$. Moreover, $\tilde{\tau}$ is an isometry and hence an automorphism of the uniformity N on P.

Proof: Because τ is a lattice automorphism of R, $\tilde{\tau}$ maps minimal Cauchy prefilters on minimal Cauchy prefilters. Clearly $\tilde{\tau}$ is bijective. In order to show "$\tilde{\tau}$ is an isometry" consider the following equivalent assertions:

$(x,y) \in N_a$

$\exists\ \sigma \in T$ such that $\sigma a \in x, y$

$\exists\ \sigma' \in T$ such that $\tau^{-1} \sigma' a \in x, y$

$\exists\ \sigma' \in T$ such that $\sigma' a \in \tilde{\tau} x, \tilde{\tau} y,$

$(\tau\tilde{x}, \tau\tilde{y}) \in N_a$. □

(343) Definition: $\tilde{T} \underset{\text{def}}{=} \{\tilde{\tau} \mid \tau \in T\}$.

(344) Proposition:

(i) Let be $\tau \in T$, $a \in R$. Then $\tilde{\tau}[\tilde{a}] = \widetilde{\tau a}$.

(ii) $\sim : \tau \mapsto \tilde{\tau}$ is a homomorphism from T into the group of isometries of P.

(iii) Let be K the kernel of the homomorphism $\sim$ and $\tilde{\mathcal{T}}$ the topology on $\tilde{T}$, which is transferred to $\tilde{T}$ by the algebraic isomorphism $\tilde{T} \cong T/K$. Then $\tilde{\mathcal{T}}$ is Hausdorff and possesses a subbase of its $Id_{\bar{P}}$ neighbourhood filter consisting of sets of the form

$$\widetilde{T(a)} = \tilde{T}(\tilde{a}) \underset{\text{def}}{=} \{\tilde{\tau} \in \tilde{T} \mid \tilde{\tau}[\tilde{a}] \cap \tilde{a} \neq \emptyset\}, \quad a \in R_o.$$

Proof:

(i) $\tilde{\tau}[\tilde{a}] = \{\tilde{\tau}x \mid x \in \tilde{a}\}$

$= \{\tilde{\tau}x \mid a \quad x\}$

$= \{\tilde{\tau}x \mid \tau a \in \tilde{\tau}x\}$

$= \{y \mid \tau a \in y\} = \tau a.$

(ii) $\tau \circ \sigma \ x = \{\tau\sigma a \mid a \in x\}$

$= \tilde{\tau}\{\sigma a \mid a \in x\}$

$= \tilde{\tau} \circ \tilde{\sigma}\ x.$

(iii) T/K is Hausdorff iff K is closed ([BGT] III § 2.6 Prop. 18 a).

Let us assume: $\exists\ \sigma \in \bar{K}$ but $\sigma \notin K$. It follows that:

$\sigma \notin K$

$\tilde{\sigma} \neq Id_P$

$\exists\ x \in P$ such that $\tilde{\sigma}\ x \neq x$

$\exists\ a \in R_o$ such that $N_a(\tilde{\sigma}x) \cap N_a(x) = \emptyset$, since P is a Hausdorff space. Now choose

$b \in R_o$ such that $N_b^2 \subset N_a$ and $\varphi \in T$ such that $\varphi\, b \in x$. Since $\sigma \in \bar{K}$, there exists a $\tau \in \sigma\, T(\varphi b) \cap K \neq \emptyset$, which implies:

$\sigma\,\varphi\, b \wedge \tau\,\varphi\, b \neq 0$. From $\varphi\, b \in x$ we infer $\sigma\,\varphi\, b \in \tilde{\sigma}\, x$ and

$\tilde{\sigma}\, x \in \widetilde{\sigma\,\varphi\, b}$. Analogously:

$\tilde{\tau}\, x \in \widetilde{\tau\,\varphi\, b}$, further:

$\exists\, z \in \widetilde{\sigma\,\varphi\, b} \cap \widetilde{\tau\,\varphi\, b} \neq \emptyset$ (see above).

That means:

$(\tilde{\sigma}x, z) \in N_b$ and $(z, \tilde{\tau}x) \in N_b$, hence $(\tilde{\sigma}x, \tilde{\tau}x) \in N_a$ in contradiction to

$\tilde{\tau}\, x = x$ and $N_a(\tilde{\sigma}x) \cap N_a(x) = \emptyset$.

Now $\tilde{T} \cong T/K$ is made into a topological group, such that the sets $\widetilde{T(a)}$, $a \in R_o$, form a subbase for the Id_P-neighbourhood filter, as follows from standard theorems ([BGT] III § 2.6 Prop. 17). It remains to show

$\tilde{T}(\tilde{a}) = \widetilde{T(a)}$:

$$\widetilde{T(a)} = \{\tilde{\tau} \mid \tau a \wedge a \neq 0\}$$

$$= \{\tilde{\tau} \mid \widetilde{\tau a} \cap \tilde{a} \neq \emptyset\}$$

$$\{\tilde{\tau} \mid \tilde{\tau}[\tilde{a}] \cap \tilde{a} \neq \emptyset\} = \tilde{T}(\tilde{a}). \qquad \square$$

(345) Proposition: $\tilde{T}$ operates continuously on P.

Proof: We have to prove the continuity of $(\tilde{\varphi}, x) \mapsto \tilde{\varphi}\, x$. Let be given any neighbourhood $N_a(\tilde{\varphi}x)$, $a \in R_o$, of $\tilde{\varphi}\, x$. Choose $b \in R_o$ such that $N_b^2 \subset N_a$ and $\tau \in T$ such that $\tau\, b \in x$. Hence

$U = \tilde{\varphi}\, \widetilde{T(\tau b)} \times \widetilde{\tau b}$ is a neighbourhood of $(\tilde{\varphi}, x)$. Assume $(\tilde{\psi}, y) \in U$.

We want to show: $\tilde{\psi}\, y \in N_a(\tilde{\varphi}x)$. It follows:

$\exists\, z \in \tilde{\varphi}\,[\widetilde{\tau b}] \cap \tilde{\psi}[\widetilde{\tau b}] \neq \emptyset$

$\tilde{\varphi}\, x \in \tilde{\varphi}[\widetilde{\tau b}]$, $\tilde{\psi}\, y \in \tilde{\psi}[\widetilde{\tau b}]$ since $x, y \in \widetilde{\tau b}$

$(\tilde{\varphi}x, z) \in N_b$ and $(z, \tilde{\psi}y) \in N_b \Rightarrow (\tilde{\varphi}x, \tilde{\psi}y) \in N_b^2 \subset N_a$. $\square$

(346) Proposition: $\tilde{T}$ operates "almost transitively" on P, that means: for any $x \in P$, $\tilde{T}\,x$ is dense in P.

Proof: Let be $y \in P$ and $\tilde{a} \in \tilde{R}$ such that $y \in \tilde{a}$ any neighbourhood of y. It follows:

x Cauchy prefilter

$\exists\ \tau \in T$ such that $\tau\, a \in x$

$x \in \tilde{\tau}[\tilde{a}]$

$\tilde{\tau}^{-1}\ x \in \tilde{a}$

$\tilde{T}\,x \cap \tilde{a} \neq \emptyset$. □

(347) Proposition:

(i) The topology $\mathfrak{T}$ on $\tilde{T}$ coincides with the topology of compact convergence $\mathfrak{c}$.

(ii) $\tilde{T}$ is uniformly equicontinuous (see [BGT] X § 2.2 Def. 2).

(iii) The uniformity of compact convergence $\mathfrak{c}^u$ coincides on $\tilde{T}$ with the uniformity of pointwise convergence $\mathfrak{s}^u$ (see [BGT] X § 1.3).

Proof:

(i) The uniformity of compact convergence is generated by the fundamental system of entourages of the form

$\tilde{T}(K,N_b) \underset{def}{=} \{(\varphi,\psi) \in \tilde{T}\times\tilde{T} \mid \forall x \in K, (\varphi x, \psi x) \in N_b\}$, $K \subset P$ compact and $b \in R_o$.

<u>$\mathfrak{T}$ finer than $\mathfrak{c}$:</u> Consider any neighbourhood $\tilde{T}(K,N_b)(\varphi)$ of $\varphi \in \tilde{T}$. Choose $a \in R_o$ as a kernel of b and

$K \subset \bigcup_{i=1\ldots n} \widetilde{\tau_i a}$ as a finite covering of the compact set K,

$\tau_i \in T$. Take $U = \bigcap_{i=1\ldots n} \widetilde{T(\tau_i a)}$ as an Id_P-neighbourhood w.r. to $\mathfrak{T}$. We will show:

$\varphi \circ U \subset \tilde{T}(K,N_b)(\varphi)$. Let be $\psi \in \varphi \circ U$.

It follows:

$\forall\ x \in K\ \exists\ j \in \{1...n\}$ such that $x \in \widetilde{\tau_j a}$ and

$\varphi[\widetilde{\tau_j a}] \cap \psi[\widetilde{\tau_j a}] \neq \emptyset$. Now $\varphi[\widetilde{\tau_j a}]$ is a kernel of $\varphi[\widetilde{\tau_j b}]$, hence

$\varphi x,\ \psi x \in \varphi\,\widetilde{\tau_j b}$ and $(\varphi x, \psi x) \in N_b$.

This means:

$\forall\ x \in K,\ (\varphi x, \psi x) \in N_b$

$(\varphi, \psi) \in \tilde{T}(K,N_b)$

$\psi \in \tilde{T}(K,N_b)(\varphi)$.

<u>$\mathfrak{T}$ coarser than $\mathfrak{c}$:</u> It is sufficient to consider a given φ-neighbourhood w.r. to $\mathfrak{T}$ of the form $\varphi \circ T(a)$, $a \in R_o$. Choose some $x \in \tilde{a}$ and $b \in R_o$ such that $N_b(x) \subset \tilde{a}$. This works because $\tilde{a}$ is open. Now consider the compact set $\bar{\tilde{a}}$ and the corresponding φ-neighbourhood $M(\varphi) \underset{def}{=} \tilde{T}(\bar{\tilde{a}},N_b)(\varphi)$ w.r. to $\mathfrak{c}$. We will show: $M(\varphi) \subset \varphi \circ \widetilde{T(a)}$. $\psi \in M(\varphi)$ implies:

$\forall\ y \in \bar{\tilde{a}}\ \exists\ \tau \in T$ such that $\varphi y,\ \psi y \in \widetilde{\tau b}$ and further:

$\varphi x,\ \psi y \in \tilde{\tau}[\tilde{a}]$

$x,\ \varphi^{-1}\psi x \in \varphi^{-1}\ \tilde{\tau}[\tilde{b}]$

$\varphi^{-1}\psi x \in N_b(x)$

$x,\ \varphi^{-1}\psi x \in \tilde{a}$

$\psi x \in \varphi[\tilde{a}],\ \psi[\tilde{a}]$

$\varphi[\tilde{a}] \cap \psi[\tilde{a}] \neq \emptyset$

$\psi \in \varphi \circ \widetilde{T(a)}$.

(ii) $\tilde{T}$ consists of isometries.

(iii) Follows from (ii) by [BGT] X § 2.4 Th. 1. □

3.5 CHAINS I

The use of "chains" to measure the distance between two points has been suggested by several authors. We only mention G. Ludwig [LUD 2] and W. Büchel (in [BOE]).

A chain joins two points and consists of a sequence of congruent, overlapping regions.

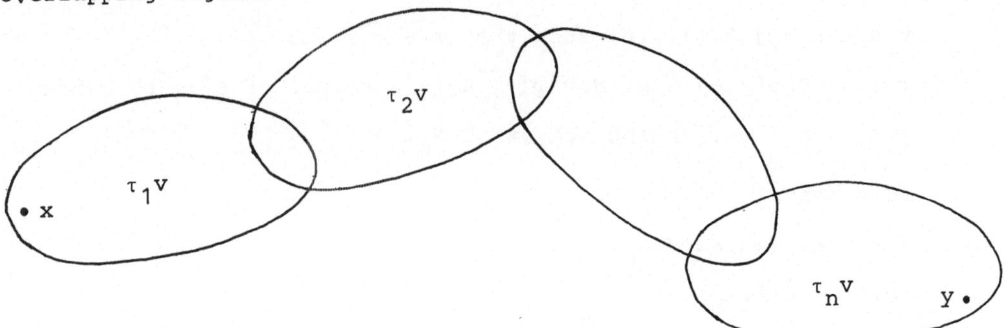

fig. (351)

The precise definition reads as follows.

(352) Definition:

(i) Let be $x, y, x', y' \in P$; $v, v' \in R_o$;

$\tau_1, \ldots \tau_n, \tau'_1, \ldots \tau'_{n'} \in T$.

$(x,y,v,\tau_1,\ldots\tau_n) \cong (x',y',v',\tau'_1,\ldots\tau'_{n'}) \underset{def}{\Leftrightarrow}$

$x = x'$, $y = y'$, $n = n'$ and $\tau_i\ v = \tau'_i\ v'$ for $i=1\ldots n$.

Clearly, $\cong$ is an equivalence relation.

(ii) A <u>chain</u> between $x \in P$ and $y \in P$ of order $v \in R_o$ of length $n \in \mathbb{N}$ is defined as an $\cong$- equivalence class of $(n+3)$-tuples of the form $(x,y,v,\tau_1,\ldots\tau_n)$ such that:

$x \in \tau_1 v$, $y \in \tau_n v$ and

$\tau_i\ v \wedge \tau_{i+1}\ v \neq 0$ for all $i=1\ldots n-1$.

The equivalence class will be denoted by

$[x,y,v,\tau_1\ldots\tau_n]$.

If $[x,y,v,\tau_1\ldots\tau_n]$ is a chain, there exist points $x_i \in \tau_i v \cap \tau_{i+1} v$, $i=1\ldots n-1$. Hence $(x_o=x, x_1,\ldots x_{n-1}, x_n=y)$ is an N_v-chain in terms of [BGT] II § 4.4, Def. 3, that is:

$(x_i, x_{i+1}) \in N_v$ for $i=0,\ldots n-1$.

The existence of chains is not insured. If, for instance, P is not connected, there may be no chains at all between x and y of order v. Therefore we need to assume a further axiom of connectedness. It is physically plausible, that not only the space P but in some sense even T should be connected.

Taking into account that, up to now, T is not necessarily complete, it would be unreasonable to assume connectedness but rather we assume a "pre-connectedness". That would mean, any two transport mappings τ, $\sigma \in T$ would be connectable by an "N-chain", where N is some entourage w.r. to the left uniformity $\mathcal{t}^s$ on T:

$$N = \{(\varphi,\psi) \in T\times T \mid \varphi^{-1}\psi \in \bigcap_{i=1\dots n} T(a_i)\}$$

$$= \{(\varphi,\psi) \in T\times T \mid \forall i \in \{1\dots n\}, \psi a_i \wedge \varphi a_i \neq O\},$$

where $n \in \mathbb{N}$, $a_i \in R_o$ for $i=1\dots n$.
Clearly it suffices to take $\tau = Id_R$.

(353) <u>Axiom R5:</u> $\forall\, \sigma \in T \;\; \forall\, n \in \mathbb{N}$

$\forall\, a_1,\dots a_n \in R_o \;\exists\, m \in \mathbb{N}$

$\exists\, \tau_1,\dots \tau_m \in T$ such that

$\tau_1 = Id_R,\; \tau_m = \sigma$ and

$\forall\, i=1\dots n \;\forall\, j=1\dots m,\; \tau_{j+1}\, a_i \wedge \tau_j\, a_i \neq O.$

"Every transport can be composed from small transports".

From this (and from the construction of transport mappings in section 2) it is clear that, for instance, reflections are not considered as transport mappings, although they are isometries.

(354) Proposition: Axiom R5 is equivalent to the following: Let be U any Id_R-neighbourhood in T, then $T = \bigcup_{n\in\mathbb{N}} U^n$ holds; that is, T is generated by arbitrarily small Id_R-neighbourhoods.

Proof:

1. Assume R5 holds. Let be $V = \bigcap_{i=1\ldots n} T(a_i)$ and $\sigma \in T$. It follows:

   $\tau_2 \in V,\ \tau_3^{-1}\ \tau_3 \in V, \ldots\ \tau_{m-1}^{-1}\ \sigma \in V$, and, since

   $\sigma = \tau_2 \circ (\tau_2^{-1}\tau_3) \circ (\tau_3^{-1}\tau_4) \circ \ldots (\tau_{m-1}^{-1}\sigma):\quad \sigma \in V^{m-1}$.

2. Conversely, let $a_1, \ldots a_n \in R_o$ and $\sigma \in T$ be given. Define $U = \bigcap_{i=1\ldots n} T(a_i)$. Because of $T = \bigcup_{n\in\mathbb{N}} U^n$, σ can be written as $\sigma = \sigma_2\ \sigma_3 \ldots \sigma_m$ where $\sigma_i \in U$ for i=2...m.

   Now define

   $\tau_1 = Id_R$ and $\tau_{j+1} = \tau_j\ \sigma_{j+1}$ for j=1...m-1.

   It follows:

   $\sigma = \sigma_2\ \sigma_3 \ldots \sigma_m = (\tau_1^{-1}\tau_2)(\tau_2^{-1}\tau_3)\ldots(\tau_{m-1}^{-1}\tau_m) = \tau_m$ and

   $\tau_j^{-1}\ \tau_{j+1} = \sigma_{j+1} \in U$, that is:

   $\forall\ i=1\ldots n\ \forall\ j=1\ldots m-1,\ \tau_{j+1}\ a_i \wedge \tau_j\ a_i \neq 0$. □

(355) Proposition: Let be $x, y \in P$, $v \in R_o$. Then there exist chains between x and y of order v.

Proof: Since x, y are Cauchy prefilters there exist $\tau, \rho \in T$ such that $x \in \tau v$, $y \in \rho v$. Now apply axiom R5 for n = 1, $a_1 = v$ and $\sigma = \tau^{-1}\ \rho$. This yields $\tau_j \in T$, j=1...m, such that $\tau_{j+1}\ v \wedge \tau_j\ v \neq 0$ and $\tau_1 = Id_R$, $\tau_m = \sigma$. Hence $[x,y,v,\tau\tau_1,\tau\tau_2,\ldots\tau\tau_m]$ is the desired chain. □

(356) Definition: The length of chains of order v between x and

y is bounded from below by 1. Therefore there are chains which attain the minimal length $n_{min} \underset{def}{=} \lambda(x,y,v)$. We shall call them minimal chains.

(357) Lemma: $v < w \Rightarrow \lambda(x,y,v) \geq \lambda(x,y,w)$

Proof: Let be $[x,y,v,\tau_1 \ldots \tau_n]$ a minimal chain. It follows, that $[x,y,w,\tau_1 \ldots \tau_n]$ is a chain, hence $\lambda(x,y,w) \leq n$. □

(358) Lemma: Let be $x, y \in P$ such that $x \neq y$.
Then there exists a $v \in R_o$ such that $\lambda(x,y,v) \geq 2$.

Proof: Because P is Hausdorff, there exists a $v \in R_o$ such that $y \notin N_v(x)$. Hence there is no chain between x and y of order v and length 1. □

(359) Proposition: Let be $x, y \in P$ and $x \neq y$. The set of integers $\{\lambda(x,y,v) \mid v \in R_o\}$ is unbounded.

Proof: Assume $\lambda(x,y,v) \leq n$. Consider some arbitrary $a_o \in R_o$ and further regions $a_1 \in R_o$ such that $N^2_{a_1} \subset N_{a_o}$,

$a_2 \in R_o$ such that $N^4_{a_2} \subset N^2_{a_1} \subset N_{a_o}$, etc. ... until

$a_k \in R_o$ such that $N^{(2^k)}_{a_k} \subset \ldots \subset N_{a_o}$ and $2^k \geq n$.

According to (355) there exists a chain $[x,y,a_k,\tau_1^{(k)},\ldots\tau_{l_o}^{(k)}]$. From this chain we will construct an a_{k-1}-chain in the following manner:

Let be $x_2 \in \widetilde{\tau_2^k a_k} \cap \widetilde{\tau_3^k a_k} \neq \emptyset$, hence $x_2 \in N^2_{a_k}(x) \subset N_{a_{k-1}}(x)$

$\exists\ \tau_1^{(k-1)} \in T$ such that $x, x_2 \in \widetilde{\tau_1^{(k-1)} a_{k-1}}$,

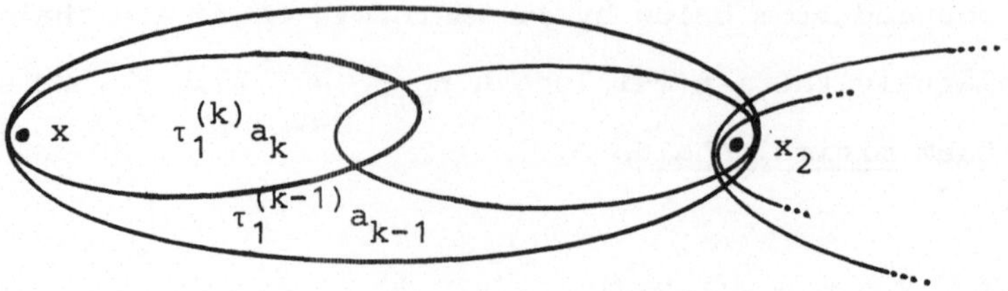

fig. (3510)

etc., until we obtain a chain

$[x,y,a_{k-1},\tau_1^{(k-1)},\ldots\tau_{l_1}^{(k-1)}]$ such that $l_1 = \left[\frac{l_o+1}{2}\right]$

($[\ldots]$ denotes the greatest integer function).

From $l_o \leq n \leq 2^k$ we obtain $l_1 \leq 2^{k-1}$. We proceed recursively, constructing chains of order a_{k-2}, a_{k-3},... until a_{k-k} is reached. The last chain $[x,y,a_{k-k},\tau_1^{(k-k)},\ldots\tau_{l_k}^{(k-k)}]$ has length $l_k \leq 2^{k-k} = 1$. Hence $\lambda(x,y,a_o) = 1$ in contradiction to (358). □

(3511) Proposition: Let be $K \subset P$ a precompact subset, $x \in P$, $v \in R_o$. Then the two sets of integers $\{\lambda(x,y,v) \mid y \in K\}$ and $\{\lambda(z,y,v) \mid z,y \in K\}$ are bounded.

Proof (from [TIT] p. 196, 3.4): Cover K with finitely many $\widetilde{\tau_\nu v}$, $\tau_\nu \in T$, choose points $y_\nu \in \widetilde{\tau_\nu v}$ and notice:

$\forall\, y \in \widetilde{\tau_\nu v},\ \lambda(x,y,v) \leq \lambda(x,y_\nu,v) + 1$ and

$\forall\, z \in \widetilde{\tau_\nu v}\ \forall\, y \in \widetilde{\tau_\mu v},\ \lambda(z,y,v) \leq \lambda(y_\nu,y_\mu,v) + 2.$ □

(3512) Proposition: Let be $K \subset P$ precompact.

$\exists\, v \in R_o\ \forall\, a \in R_o,\ a < v \Rightarrow \forall\, n \in \mathbb{N},\quad N_a^n(K)$ is precompact.

Proof (see [FRE] 1.3, 1.4): We will show first, that $N_a(K)$ is precompact. Let be $u \in R_o$, v a kernel of u and $K \subset \bigcup_{\nu=1\ldots n} \widetilde{\tau_\nu v}$, $\tau_\nu \in T$ a covering of K, $a \in R_o$, $a < v$ and $x \in N_a(K)$, that is:

$x \in \widetilde{\tau a}$ and some $y \in \widetilde{\tau a} \cap K$, $\tau \in T$.

It follows:

$\exists\ \nu \in \{1\ldots n\}$ such that $y \in \widetilde{\tau_\nu v}$

$\widetilde{\tau a} \subset \widetilde{\tau v}$

$y \in \widetilde{\tau v} \cap \widetilde{\tau_\nu v} \neq \emptyset$

$\tau_\nu^{-1}\ \tau\ v \wedge v \neq 0$

$\tau_\nu^{-1}\ \tau\ v < u$

$\tau v < \tau_\nu u$

$x \in \tau_\nu u.$

This shows: $N_a(K) \subset \bigcup_{\nu=1\ldots n} \widetilde{\tau_\nu u}$

Now, $N_a(K)$ is a subset of a finite union of precompact sets, hence precompact.

The precompactness of $N_a^n(K)$ follows by induction. □

The last propositions remain valid if one substitutes "relatively compact" for "precompact", since the two concepts coincide in complete spaces. For the next section we need the following

(3513) Lemma: Let be $K, L \subset P$ precompact.

Then $H \underset{def}{=} \bigcup\{\varphi[L] \mid \varphi \in \tilde{T}, \varphi[L] \cap K \neq \emptyset\}$ is precompact.

Proof: Choose $v \in R_o$ according to (3512) and let be $n = \max\{\lambda(x,y,v) \mid x,y \in L\}$ (this exists by (3511). Consider any $x \in \varphi[L] \cap K \neq \emptyset$ and $y \in \varphi[L]$. $\varphi^{-1}x$ and $\varphi^{-1}y$ are points in L and can be connected with a v-chain of length $k \leq n$. The latter is also true for x and y. This shows $y \in N_V^k(K)$ and, therefore, $H \subset N_V^n(K)$. As a subset of a precompact set (by (3512)), H is precompact. □

The length of minimal chains $\lambda(x,y,a)$ is an approximate measure of the distance between x and y "in units of a". It can be made more accurate if the region $a \in R_o$ is scaled down, but then $\lambda(x,y,a)$ increases. To overcome this difficulty, let us fix a pair $(u,v) \in P \times P$ of distinct points as "unit length" and consider the "chain quotient" $\frac{\lambda(x,y,a)}{\lambda(u,v,a)}$ as a measure of the distance, which becomes more and more accurate, the smaller a is chosen.

This can be made precise in the following manner. Note that $\lambda(x,y,a)$ depends only on the congruence class of a, $[a] \in R_o/T$, hence the chain quotient may be considered as a real function f on R_o/T, $[a] \mapsto f[a] = \frac{\lambda(x,y,a)}{\lambda(u,v,a)}$, for fixed $x,y,u,v \in P$. Now R_o/T is a (downwards) <u>directed set</u> w.r. to the partial ordering defined by

$$[a] \subset [b] \underset{\text{def}}{=} \exists\, a_1 \in [a],\ b_1 \in [b] \text{ such that } a_1 < b_1.$$

This can be easily proved by using (332). Hence it is meaningful to speak of the <u>limit</u> of the function f in the sense of [BGT] I § 7.3, Def. 3, Ex. 2), which will be denoted simply by $\lim\limits_{a \in R_o} \frac{\lambda(x,y,a)}{\lambda(u,v,a)}$. This leads to the

(3514) Definition: If for all $x,y,u,v \in P$ (such that $u \neq v$) the limit

$$\lim_{a \in R_o} \frac{\lambda(x,y,a)}{\lambda(u,v,a)} \underset{\text{def}}{=} \Lambda(x,y,u,v)$$

exists, it is called the <u>chain distance</u> (more precisely: chain distance quotient) between x and y. If u,v are fixed, we will abbreviate $\Lambda(x,y,u,v) = d(x,y)$.

3.6 COMPLETION OF THE GROUP $\tilde{T}$

In order to connect our theory with the axioms in [FRE], we would have to show, for instance, transitivity and completeness (w.r. to c^u) of the group of congruent mappings $\tilde{T}$. Of course, $\tilde{T}$ is not necessarily complete, and it is thus a question of an appropriate completion of $\tilde{T}$ and of proving transitivity and the other axioms in [FRE] for this completion (if possible at all).

Now the difficulty emerges that a completion of $\tilde{T}$ w.r. to the uniformity of compact convergence is not possible in all cases. For example, the group $GL(\mathbb{R},n)$ is c-dense in the set of all matrices $M_n(\mathbb{R})$, as mentioned in [BGT] X § 3.5.
Another counter-example is the discrete space $\mathbb{N}$, endowed with the metric

$$d(n,m) = \begin{cases} 1 & \text{if } n \neq m \\ 0 & \text{if } n = m \end{cases}$$

and the group of isometries of $\mathbb{N}$ ([BGT] X § 3 Ex. 19c). This group together with the lattice of finite subsets of $\mathbb{N}$ satisfies the axioms R1 to R4.

On the other hand, if we choose another uniformity for the completion of $\tilde{T}$, some extra work will be necessary to prove several theorems in [FRE] under the modified conditions. Let us proceed in this way and consider the topology c ($=s=\tilde{\tau}$ by (347)) on $\tilde{T}$ and the corresponding left (resp. right) uniformity c^s ($=c^u$) (resp. c^d) and the supremum of the two $c^z = c^s \vee c^d$, the two-sided uniformity. Further, the space of all mappings $P \to P$, endowed with the uniformity of pointwise convergence, will be denoted by $F_s(P,P)$.

(361) Proposition: Each $\mathcal{c}^z$-Cauchy-filter on $\tilde{T}$ converges in the space $F_s(P,P)$. The set of limit points $\bar{T} \subset F_s(P,P)$ is a uniformly equicontinuous group of homeomorphisms of P. $\bar{T}$ is s^z-complete and $\tilde{T}$ is s-dense in $\bar{T}$.

Proof: [BGT] X § 3 Ex. 19d) □

(362) Corollary: $\bar{T}$ is $\mathcal{c}^z$-complete, $\tilde{T}$ is $\mathcal{c}$-dense in $\bar{T}$. Hence $\bar{T}$ may be identified with the $\mathcal{c}^z$-completion of $\tilde{T}$. The mappings $\tau \in \bar{T}$, which also will be called "congruent mappings", are automorphisms of the uniformity N on P.

Proof: Follows from [BGT] X § 2.4 Th. 1 □

(363) Definition: In accordance with (344)(iii) we define

$$\tilde{T}(A,B) \underset{\text{def}}{=} \{\tau\in\tilde{T} \mid A\cap\tau[B]\neq\emptyset\},$$

$$\tilde{T}(x,B) \underset{\text{def}}{=} \tilde{T}(\{x\},B),$$

$$\tilde{T}(A) \underset{\text{def}}{=} \tilde{T}(A,A).$$

Analogously: $\bar{T}(A,B)$, $\bar{T}(x,B)$, $\bar{T}(A)$.

(364) Proposition:

(i) The tripel $(\bar{R},\subset,\bar{T})$ satisfies the axioms R1 to R5. The corresponding space of points $\bar{P}$ is canonically isomorphic to P as a uniform space. (The bar denotes in sequel a transfer of a definition within the $(R,<,T)$-theory onto the "model" $(\bar{R},\subset,\bar{T})$.)

(ii) The sets $\bar{T}(A)$, $A \in \bar{R}_o$, form a subbase of the topology $\bar{\mathcal{t}}$ on $\bar{T}$, which coincides with the topology of compact convergence $\mathcal{c}$.

Hence to each proposition proved with R1 to R5 there corresponds a new proposition (the "bar" version) which is valid for $(\bar{R},\subset,\bar{T})$.

Proof: (i)

1. $(\bar{R},\subset)$ is a distributive lattice with least element $\emptyset$. $\bar{R}$ is not empty, since $P \neq \emptyset$.

2. $\bar{T}$ consists of homeomorphisms and thus induces a group of $(\bar{R},\subset)$-automorphisms (which we have identified with $\bar{T}$).

3. Let be $A \in \bar{R}_o$. Since A is open, there exists some $\tilde{a} \in \tilde{R}_o$ such that $\tilde{a} \subset A$. Choose $\tilde{b} \in \tilde{R}_o$ as a kernel of a kernel of $\tilde{a}$, hence $N_b^3(\tilde{b}) \subset \tilde{a}$. We will show, that $\tilde{b} \in \tilde{R} \subset \bar{R}$ is a kernel (w.r. to $\bar{T}$) of A. Thus assume $\tau \in \bar{T}$ such that $\tau[\tilde{b}] \cap \tilde{b} \neq \emptyset$, say $\tau y \in \tau[\tilde{b}] \cap \tilde{b}$.

   Further let $x \in \tilde{b}$ be arbitrary. Since $\tilde{T}$ is c-dense in $\bar{T}$, there exists a $\varphi \in \tilde{T} \cap \bar{T}(\bar{\tilde{b}},N_b)(\tau)$. This implies:

   $\forall\ z \in \bar{b}\ \exists\ \sigma \in \tilde{T}$ such that $\tau z,\ \ \varphi z \in \sigma[\tilde{b}]$

   $\tau y \in \tilde{b}$

   $\tau y,\ \ \varphi y \in \sigma_1[\tilde{b}]$

   $\varphi y \in \varphi\ [\tilde{b}]$

   $\varphi x \in \varphi\ [\tilde{b}]$

   $\varphi x,\ \tau x \in \sigma\ [\tilde{b}]$

   Hence: $\tau x \in N_b^3(\tilde{b}) \subset \tilde{a} \subset A$.

   Since $\tau x \in \tau[\tilde{b}]$ was arbitrary, we have proved $\tau[\tilde{b}] \subset A$, thereby R3.

4. Let be given $A, V \in \bar{R}_o$. There is some $\tilde{a} \in \tilde{R}_o$ such that $\tilde{a} \subset V$ and $\bar{A}$ can be covered by finitely many $\tau_\lambda[\tilde{a}]$, $\tau_\lambda \in \tilde{T}$, $\lambda=1...n$. From this the claim of R4 follows immediately.

   Notice that part(ii) of the proposition follows from proposition (347) (in the "bar" version) and may be used in the following discussion.

5. Let be $\sigma \in \bar{T}$ and U some Id_P-neighbourhood in $\bar{T}$ w.r. to $\subset$. $\tilde{T}$ being $\subset$-dense in $\bar{T}$, there exists some $\sigma' \in \tilde{T}$ such that $\sigma'^{-1}\ \sigma \in U$. Moreover, $U' \underset{def}{=} U \cap \tilde{T}$ is a Id_P-neighbourhood in $\tilde{T}$ w.r. to $\subset$. Between Id_P and σ' there exists a U'-chain by axiom R5: $(\tau_1 = Id_P, \tau_2, \ldots \tau_n = \sigma')$. Consequently, $(\tau_1, \ldots \tau_n, \sigma)$ is a U-chain.

6. $\bar{P} \cong P$.

6.1. Let F be a filter on P. The assiquement $F \mapsto F \cap \bar{R}$ maps filters on P onto prefilters in $(\bar{R}, \subset)$. We will show that the minimal Cauchy filters on P are bijectively mapped onto the minimal Cauchy prefilters in $(\bar{R}, \subset, \bar{T})$.

Let F be a Cauchy filter on P and $B \in \bar{R}_o$, $\tilde{a} \subset B$, $\tilde{a} \in \tilde{R}_o$. Since $\underset{\sim}{F}$ is a Cauchy prefilter in $(R, <, T)$ (see (336)(vii)), there exists some $\tau \in T$ such that $\tau\ a \in \underset{\sim}{F}$, that is: $\tilde{\tau}[\tilde{a}] \in F$. By the filter property of F we have $\tilde{\tau}[B] \in F \cap \bar{R}$, where $\tilde{\tau} \in \tilde{T} \subset \bar{T}$. This proves $F \cap \bar{R}$ to be Cauchy.

Conversely, let C be a Cauchy prefilter in $(\bar{R}, \subset, \bar{T})$ and F the filter on P generated by C, hence $C = F \cap \bar{R}$. In view of (336)(ix) it suffices to show that $\underset{\sim}{F} = \underset{\sim}{C}$ is a Cauchy prefilter in $(R, <, T)$. Let $a \in R_o$ be given and choose some $b \in R_o$ such that $N_b^2(\tilde{b}) \subset \tilde{a}$. The Cauchy property of C yields a $\tau \in \bar{T}$ such that $\tau[\tilde{b}] \in C$. Since $\tilde{T}$ is $\subset$-dense in $\bar{T}$, we find some $\tilde{\sigma} \in \tilde{T} \cap \bar{T}(\bar{\tilde{b}}, N_b)(\tau)$.

Like in 3. one can show $\tau[\tilde{b}] \subset \tilde{\sigma}[\tilde{a}]$, hence $\tilde{\sigma}[\tilde{a}] \in C$ and $\sigma\ a \in \underset{\sim}{C}$.

Thereby we have shown, that there exists a Cauchy filter F on P, which is mapped by $F \mapsto F \cap R$ onto a given Cauchy prefilter C in $(\bar{R}, \subset, \bar{T})$. Now we know, that minimal Cauchy filters on P are just neighbourhood filters, since P is complete. Hence $F \mapsto F \cap \bar{R}$ is injective, if restricted to minimal Cauchy filters. Since this map respects the inclusion of (pre)filters, minimal Cauchy filters on P are bijectively mapped onto minimal Cauchy prefilters in $(\bar{R},\subset,\bar{T})$.

6.2. We will identify P and $\bar{P}$ as sets according to 6.1., and consider the uniformity $\bar{N}$ on P given by entourages of the form

$\bar{N}_a = \{(x,y)\in P\times P \mid \exists\tau\in\bar{T} \text{ such that } x,y\in\tau[a]\}$, $a \in \bar{R}_o$.

It remains to show that $\bar{N}$ and N coincide as uniform structures.

6.2.1. $\bar{N}$ is finer than N.

Let N_a, $a \in R_o$, be given and choose $b \in R_o$ as a kernel of a. We will show $\bar{N}_{\tilde{b}} \subset N_a$. Thus take any $(x,y) \in \bar{N}_{\tilde{b}}$, which means: $x, y \in \sigma[\tilde{b}]$ with some $\sigma \in \bar{T}$. By part (ii), $\tilde{T}$ is $\bar{\mathfrak{t}}$-dense in $\bar{T}$. Hence there exists a $\tau \in \tilde{T}$ such that $\sigma[\tilde{b}] \cap \tau[\tilde{b}] \neq \emptyset$. By the kernel property of b, $\sigma[\tilde{b}] \subset \tau[\tilde{a}]$ follows, hence $x, y \in \tau[\tilde{a}]$ and $(x,y) \in N_a$.

6.2.2. $\bar{N}$ is coarser than N. Let $\bar{N}_a$, $a \in \bar{R}_o$, be given. Since a is open, it contains some $\tilde{b} \subset a$, $\tilde{b} \in \tilde{R}_o$. We will show $N_b \subset \bar{N}_a$. Take any $(x,y) \in N_b$, that is: $x, y \in \tau[\tilde{b}]$ with some $\tau \in \tilde{T}$. We infer $x, y \in \tau[a]$ and, since $\tilde{T} \subset \bar{T}$, $(x,y) \in N_a$. □

Because we have only the completeness of $\bar{T}$ w.r. to the uniformity C^z instead the coarser $C^u = C^s$, as H. Freudenthal postulates, we have to prove some theorems in [FRE] under our weaker conditions.

(365) Proposition: Let $K \subset P$ be compact. $\tilde{T}(K)$ is precompact w.r. to $\mathfrak{T}^z$.

Proof: According to [BGT] II § 4.2 Th. 3, it is to prove that $\tilde{T}(K)$ can be covered by finitely many sets, which are $\mathfrak{T}^z$-small.

1. Let $N = N_s \cap N_d$ be any $\mathfrak{T}^z$-entourage, which may assumed to be of the form

$N_s = \{(\tilde{\varphi},\tilde{\psi}) \in \tilde{T}\times\tilde{T} \mid \forall i=1\ldots l, \varphi^{-1}\psi a_i \wedge a_i \neq 0\}$,

$N_d = \{(\tilde{\varphi},\tilde{\psi}) \in \tilde{T}\times\tilde{T} \mid \forall i=1\ldots l, \varphi\psi^{-1} a_i \wedge a_i \neq 0\}$,

where $a_i \in R_o$, $i=1\ldots l$. Because of $K' \supset K \Rightarrow \tilde{T}(K') \supset \tilde{T}(K)$ we may suppose without loss of generality, that K contains all $\tilde{a}_i$. By lemma (3513) there exists a compact subset $\bar{H} \subset P$ such that

$$\bigcup_{\varphi \in \tilde{T}(K)} \varphi[K] \subset \bar{H}.$$

Let $c_i \in R_o$ be kernels of a_i and

$$c \underset{def}{=} \bigtimes_{i=1\ldots l} \tilde{c}_i \subset W \underset{def}{=} \bar{H} \times \bar{H} \times \ldots \times \bar{H}.$$

W is compact as a cartesian product of compact sets and $\tilde{T}$ operates on $P \times P \times \ldots \times P$ by means of

$$\tau(x_1,x_2,\ldots x_l) \underset{def}{=} (\tau x_1,\tau x_2,\ldots \tau x_l) \text{ for } \tau \in \tilde{T}.$$

The open covering $W \subset \bigcup_{\tau \in \tilde{T}} \tau[c]$ contains a finite covering

$$W \subset \bigcup_{\mu=1\ldots m} \tau_\mu[c].$$

2. Now we are prepared to cover $\tilde{T}(K)$. Let be $\psi \in \tilde{T}(K)$.

2.1. Because of $\tilde{c}_i \subset K$ we have $\psi[\tilde{c}_i] \subset \bar{H}$ for all $i=1\ldots l$ and therefore $\bigtimes_{i=1\ldots l} \psi[\tilde{c}_i] \subset W$. Hence $\bigtimes_{i=1\ldots l} \psi[\tilde{c}_i]$ meets some $\tau_\mu[c]$ of the above covering and we conclude:

$$\begin{aligned} \emptyset \neq & \bigtimes_{i=1\ldots l} \psi[\tilde{c}_i] \cap \tau_\mu[c] \\ = & \bigtimes_{i=1\ldots l} \psi[\tilde{c}_i] \cap \bigtimes_{i=1\ldots l} \tau_\mu[\tilde{c}_i] \\ = & \bigtimes_{i=1\ldots l} \psi[\tilde{c}_i] \cap \tau_\mu[\tilde{c}_i], \text{ further} \end{aligned}$$

$$\exists \mu \in \{1\ldots m\}\ \forall i \in \{1\ldots l\},\ \emptyset \neq \psi[\tilde{c}_i] \cap \tau_\mu[\tilde{c}_i], \Rightarrow \tau_\mu^{-1}\psi \in \bigcap_{i=1\ldots l} \tilde{T}(\tilde{c}_i) \Rightarrow$$

$$\psi \in \bigcup_{\mu=1\ldots m} \tau_\mu \circ [\bigcap_{i=1\ldots l} \tilde{T}(\tilde{c}_i)] \underset{def}{=} \bigcup_\mu T_\mu.$$

2.2. $\psi \in \tilde{T}(K) \Rightarrow \psi^{-1} \in \tilde{T}(K)$. We may repeat the argument of 2.1. and conclude analogously:

$\exists\ \nu \in \{1\ldots m\}\ \forall\ i \in \{1\ldots l\},\ \emptyset \neq \psi^{-1}[\tilde{c}_i] \cap \sigma_\nu[\tilde{c}_i]$, where

$\sigma_\nu \in \tilde{T}$ for $\nu=1\ldots m$. If follows:

$$\psi \in \bigcup_{\nu=1\ldots m} \left[\bigcap_{i=1\ldots l} \tilde{T}(\tilde{c}_i) \circ \sigma_\nu^{-1} \right] \underset{\text{def}}{=} \bigcup_\nu S_\nu .$$

2.3. Because $\psi \in \tilde{T}(K)$ was arbitrary, we have constructed a finite covering

$$\tilde{T}(K) \subset \bigcup_{\nu,\mu} S_\mu \cap T_\mu .$$

3. It remains to show:

$(S_\nu \cap T_\mu) \times (S_\nu \cap T_\mu) \subset N_s \cap N_d = N.$

Let be $\varphi, \psi \in S_\nu \cap T_\mu$. As elements of T_μ, they are of the form $\varphi = \tau_\mu \circ \alpha$ such that $\forall\ i \in \{i\ldots l\},\ \alpha[\tilde{c}_i] \cap \tilde{c}_i \neq \emptyset,\ \psi = \tau_\mu \circ \beta$ such that $\forall\ i = \{i\ldots l\},\ \beta[\tilde{c}_i] \cap \tilde{c}_i \neq \emptyset$.

Similarly, as in the proof of (316)b) one uses the kernel properties of c_i in order to conclude:

$\tilde{c}_i \subset \alpha[\tilde{a}_i], \beta[\tilde{a}_i]$. Hence $\alpha[\tilde{a}_i] \cap \beta[\tilde{a}_i] \neq \emptyset$, which proves $(\varphi,\psi) \in N_d$. Analogously $(\varphi,\psi) \in N_s$ is inferred from $\varphi, \psi \in S_\nu$. □

(366) Corollary: $\bar{T}(K)$ is $\bar{\mathcal{t}}$-compact.

Proof: The "bar" version of (365) reads:

$\bar{T}(K)$ is $\bar{\mathcal{t}}^z$-precompact. But $\bar{T}(K)$ is $\bar{\mathcal{t}}$-closed by [BGT] III § 4.5 Th. 1a) and $\bar{T}$ is $\bar{\mathcal{t}}^z$-complete. This gives the assertion. □

By (366) we have proved essentially that $\bar{T}$ operates "properly" on P. Let us recall the corresponding

(367) Definition (see [BGT] III § 4.1 Def. 1):

A topological group G, which operates continuously on a

topological space X, is said to operate <u>properly</u>, iff the mapping

$\theta : G \times X \to X \times X$

$(\tau, x) \mapsto (x, \tau x)$

is proper, that is, if for every topological space Z the mapping $\theta \times Id_Z$ maps closed sets onto closed ones.

(368) Theorem: $\bar{T}$ operates properly on P.

Proof: Since P is locally compact, the assertion is equivalent to the compactness of $\bar{T}(K)$ for any compact $K \subset P$ (see [BGT] III § 4.5 Th. 1.c). Hence it follows from (366). □

From the cited theorem in [BGT] we further infer:

(369) Theorem: $\bar{T}$ is locally compact.

(3610) Theorem: $\bar{T}$ operates transitively on P.

Proof: Let be $x \in P$. By (346) $\tilde{T}\,x$ is dense in P, hence also $\bar{T}\,x$. But $\bar{T}\,x$ is closed, since $\bar{T}$ and $\{x\}$ are closed and $\bar{T}$ operates properly. □

(3611) Theorem: $\bar{T}$ is connected.

Proof: $\bar{T}$ is generated by any Id-neighbourhood (see (354)) and is locally compact. By [BGT] III § 4.6 prop. 14, Cor. 2, $\bar{T}$ is connected. □

(3612) Definition: Let be $x \in P$.

The group $J_x \underset{def}{=} \bar{T}(\{x\}) = \{\tau\in\bar{T} \mid \tau x=x\}$, which is a compact subgroup by (366), is called the <u>stability subgroup</u> of the point x.

We shall simply write J for J_x, if x is kept fixed.

(3613) Theorem: P is a topological homogeneous space of $\bar{T}$ (see [BGT] III § 2.5).

Proof: (3613) means, that the canonical map

$\bar{T}/J_x \to P$

$\tau J_x \mapsto \tau x$

is a homeomorphism. This follows from [BGT] III § 4.2 Prop. 4, since $\bar{T}$ operates properly and transitively. □

(3614) Theorem: P is connected and paracompact.

Proof: [BGT] III § 4.6 Prop. 13. □

(36.15) Proposition: J_x does not contain any proper invariant subgroup N of $\bar{T}$.

Proof: Let be $\nu \in N$, $y \in P$. It follows that $\exists\ \tau \in \bar{T}$ such that $y = \tau x$.

$\nu y = \nu \tau x \underset{def}{=} \tau \mu x = \tau x = y$, since $\mu \in N \subset J_x$. Hence:

$\nu = Id_P$. □

With the exception of axiom (Z) all of the axioms in [FRE] are now proved in our context.

3.7 CHAINS II

For later use, we will analyze the relation between $(R,<,T)$-chains and $(\overline{R},\subset,\overline{T})$-chains. Our aim is to verify the conjecture that no really new chain is added by completion.

Above all, it is clear, that to each $(R,<,T)$-chain there corresponds a $(\tilde{R},\subset,\tilde{T})$-chain and vice versa. In the first step we will compare $(\tilde{R},\subset,\tilde{T})$-chains with $(\tilde{R},\subset,\overline{T})$-chains.

(371) Lemma: Let be $a \in \tilde{R}_o$, $x \in P$.
$\tilde{T}(a)$ is $\overline{\mathfrak{t}}$-dense in $\overline{T}(a)$, likewise $\tilde{T}(x,a)$ in $\overline{T}(x,a)$. $\overline{T}(x,a)$ is $\overline{\mathfrak{t}}$-open.

Proof: $\overline{T}(a)$ is open by $\overline{(316)}$. Thus in any neighbourhood of a $\tau \in \overline{T}(a)$ we find some $\sigma \in \tilde{T} \cap \overline{T}(a) = \tilde{T}(a)$. For $\overline{T}(x,a)$ an analogous argument is valid. Let be $\tau \in \overline{T}(x,a)$, U some Id-neighbourhood, $b \in \tilde{R}_o$, such that $\overline{N}_b(x) \subset \tau[a]$ and c a kernel of b.
Without loss of generality we may assume $x \in c \subset b \subset \tau[a]$. There exists a $\sigma \in \tilde{T}$ such that $\tau\,\sigma^{-1} \in U \cap \overline{T}(c)$.
From $\tau\,\sigma^{-1}[c] \cap c \neq \emptyset$ follows:
$\tau\,\sigma^{-1}\,x \in \tau\,\sigma^{-1}\,c \subset b \subset \tau[a]$
$x \in \sigma[a]$
$\sigma \in \tilde{T}(x,a)$.
This means: $\tilde{T}(x,a)$ is dense in $\overline{T}(x,a)$.
The same reasoning also shows that the τ-neighbourhood $(\overline{T}(c))^{-1}\,\tau$ is contained in $\overline{T}(x,a)$. Hence $\overline{T}(x,a)$ is open. □

For the purpose of comparison of minimal chain lengths we will consider a formulation of the definition of $\lambda(x,y,a)$ which refers only to the group $\tilde{T}$.

(372) Definition: Let be $x, y \in P$, $x \neq y$, $a \in \tilde{R}_o$.

$\tilde{\lambda}(x,y,a) = n$ will be defined as the least integer satisfying

$$(*) \qquad \tilde{T}(x,a)\ \tilde{T}(a)^{n-1} \cap \tilde{T}(y,a) \neq \emptyset.$$

(373) Lemma: $\tilde{\lambda}(x,y,a) = \lambda(x,y,a)$.

Proof: Let be $\tau_n \in \tilde{T}(y,a)$ such that $\tau_n = \tau_1\ \varphi_2\ \varphi_3 \ldots \varphi_n$ where $\tau_1 \in \tilde{T}(x,a)$ and $\varphi_i \in \tilde{T}(a)$, i=2...n. We conclude:

$$\begin{array}{ll}
\tau_1 \in \tilde{T}(x,a) & \Rightarrow x \in \tau_1[a] \\
\varphi_2[a] \cap a \neq \emptyset,\ \tau_2 \underset{\text{def}}{=} \tau_1\ \varphi_2 & \Rightarrow \tau_2[a] \cap \tau_1[a] \neq \emptyset \\
\varphi_3[a] \cap a \neq \emptyset,\ \tau_3 \underset{\text{def}}{=} \tau_2\ \varphi_3 & \Rightarrow \tau_3[a] \cap \tau_2[a] \neq \emptyset \\
\vdots & \\
\varphi_n[a] \cap a \neq \emptyset,\ \tau_n \underset{\text{def}}{=} \tau_{n-1}\ \varphi_n & \Rightarrow \tau_n[a] \cap \tau_{n-1}[a] \neq \emptyset \\
\tau_n \in \tilde{T}(y,a) & \Rightarrow y \in \tau_n[a].
\end{array}$$

Thus each element τ_n of (372)(*) yields a chain $[x,y,a,\tau_1,\ldots\tau_n]$ and, of course, vice versa. Minimal n such that (372)(*) holds is therefore the same as minimal chain length. □

Analogously,

(374) Lemma: Each element of

$\bar{T}(x,a)\ \bar{T}(a)^{n-1} \cap \bar{T}(y,a)$

yields a $(\tilde{R},\subset,\bar{T})$-chain and vice versa.

Now we can compare chains of different type. Each $(\tilde{R},\subset,\tilde{T})$-chain is a $(\tilde{R},\subset,\bar{T})$-chain.

Hence $\bar{\lambda} \underset{\text{def}}{=} \lambda^{(\tilde{R},\subset,\bar{T})}(x,y,a) \leq \tilde{\lambda} \underset{\text{def}}{=} \lambda^{(\tilde{R},\subset,\tilde{T})}(x,y,a)$.

Now consider some $(\tilde{R},\subset,\bar{T})$-chain $[x,y,a,\tau_1,\ldots\tau_n]$. As in the proof of (373) we have

$\tau_n = \tau_1\,\varphi_2\,\ldots\,\varphi_n,\ \tau_1 \in \bar{T}(x,a),\ \tau_n \in \bar{T}(y,a),\ \ \varphi_i \in \bar{T}(a),\ \ i=2\ldots n.$
There exists a τ_n-neighbourhood $U_n \subset \bar{T}(y,a)$, since $\bar{T}(y,a)$ is open (371). The multiplication in $\bar{T}$ is continuous, hence there exist neighbourhoods V_1 of τ_1, V_i of φ_i, $i=2\ldots n$, such that for all $\tau_1' \in V_1,\ \varphi_i' \in V_i,\ \ i=2\ldots n\colon\ \tau_n' \underset{def}{:=} \tau_1'\,\varphi_2'\,\ldots\,\varphi_n' \in U_n$. Using (371) we choose $\tau_1' \in V_1 \cap \tilde{T}(x,a),\ \varphi_i' \in V_i \cap \tilde{T}(a),\ \ i=2\ldots n$, and conclude $\tau_n' \in U_n \cap \tilde{T} \subset \tilde{T}(y,a)$. This yields a $(\tilde{R},\subset,\tilde{T})$-chain $[x,y,a,\tau_1'\ldots\tau_n']$ of the same length n. Hence $\tilde{\lambda} \le \bar{\lambda}$ and we have proved the

(375) Proposition: The completion of $\tilde{T}$ to $\bar{T}$ does not alter the length of minimal chains.

In the next step we will compare $(\tilde{R},\subset,\bar{T})$-chains with $(\bar{R},\subset,\bar{T})$-chains.

(376) Lemma: Let U be an open Id-neighbourhood w.r. to the topology $\bar{\mathcal{t}}$ on $\bar{T}$, $K \subset P$ compact. Then there exists an $e \in \tilde{P}_o$, such that for each $x \in K$, U operates transitively on $N_e(x)$.

Proof: Let V be an open Id-neighbourhood satisfying $VV^{-1} \subset U$. Let us assume the negation of the above claim:
$\forall\ e \in \tilde{R}_o\ \exists\ x(e) \in K\ \exists\ \sigma \in \tilde{T}$ such that $x(e) \in \sigma[e]$ but $\sigma[e] \not\subset U\,x(e)$.
Clearly x(e) depends only on the congruence class $[e] \in [\tilde{R}_o] \underset{def}{=} \tilde{R}_o/\tilde{T}$. Notice that $[\tilde{R}_o]$ is directed downwards in a natural way. Since K is compact, there exists a filter F on $[\tilde{R}_o]$, which is finer than the section filter, such that $x(F) \to z \in K$. Vz is open ([BGT] III § 2.5 Prop. 15), hence some $a \in \tilde{R}_o$ fulfils $N_a^2(z) \subset Vz$. $x(F) \to z \in K$ yields some $f \in \tilde{R}_o$ such that

$x(f) \in N_a(z)$, where $f \subset a$ may be assumed. The above negation implies the existence of some $y \in \sigma[f]$ such that $y \notin U\,x(f)$. Since $x(f) \in \sigma[f]$, we have

$y \in N_f(x(f)) \subset N_a(x(f)) \subset N_a^2(z) \subset Vz$.

Together with $x(f) \in Vz$ this implies $y \in VV^{-1}\, x(f) \subset U\, x(f)$ in contradiction to $y \notin U\, x(f)$. □

(377) Lemma: Consider the right uniformity $\bar{\mathcal{I}}^d$ on $\bar{T}$ and the usual uniformity $\mathcal{N}$ on P. Further let $\mathcal{P}P$ and $\mathcal{P}\bar{T}$ be equipped with the corresponding canonical uniformities, resp. topologies, likewise the subset $\bar{R}_o \subset \mathcal{P}P$. Then the two mappings

$\bar{R}_o \to \mathcal{P}\bar{T}$

(i) $a \mapsto \bar{T}(a)$,

(ii) $a \mapsto \bar{T}(x,a)$

are continuous.

Proof: (i) Let M be any $\bar{\mathcal{I}}^d$-entourage of the form $(\psi,\varphi) \in M \Leftrightarrow \psi\,\varphi^{-1} \in U$, where U is some open Id-neighbourhood w.r. to $\bar{\mathcal{I}}$. Consider the corresponding $\bar{T}(a)$-neighbourhood

$\bar{\bar{M}}(\bar{T}(a)) = \{S \in \mathcal{P}\bar{T} \mid S \subset M(\bar{T}(a)) \text{ and } \bar{T}(a) \subset M(S)\}$.

The subset $K = \overline{\bar{N}_a^2(a)}$ is compact (3512), the application of (376) yields for each $x \in K$ some neighbourhood $N_e(x)$, on which U operates transitively. Let d be a kernel of e. We may assume $d \subset a$. Consider the a-neighbourhood w.r. to the uniformity $\bar{\mathcal{N}}$ on $\bar{R}_o$ given by

$\bar{\bar{N}}_d(a) = \{b \in \bar{R}_o \mid b \subset N_d(a) \text{ and } a \subset N_d(b)\}$.

Let be $b \in \bar{\bar{N}}_d(a)$. We will show that $\bar{T}(b) \in \bar{\bar{M}}(\bar{T}(a))$. In order to prove $\bar{T}(b) \subset M(\bar{T}(a))$, let $\varphi \in \bar{T}(b)$ be given. From $b \in \varphi[b] \neq \emptyset$ follows $N_d(a) \cap N_d(\varphi[a]) \neq \emptyset$. Hence there exist $x, y, z \in P$; $\tau_1, \tau_2 \in \tilde{T}$ such that

$x \in \varphi[a] \cap \tau_1[d]$, $y \in \tau_1[d] \cap \tau_2[d]$, $z \in \tau_2[d] \cap a$.

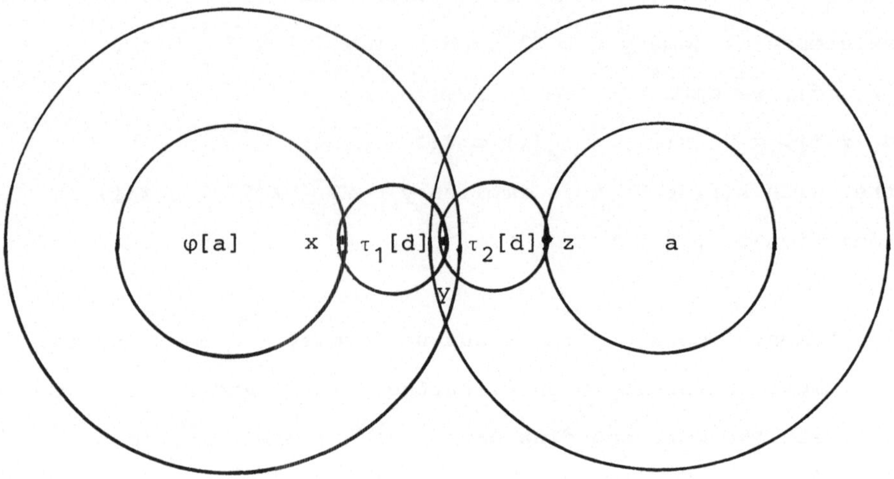

fig. (378)

We conclude $x \in N_a^2(a) \subset K$ and $z \in N_e(x)$, hence there exists a $\sigma \in U$ such that $\sigma x = z$.

It follows:

$\sigma \varphi[a] \cap a \neq \emptyset$

$\psi \underset{def}{=} \sigma \varphi \in \bar{T}(a)$ and $(\psi, \varphi) \in M$

$\varphi \in M(\bar{T}(a))$.

Thereby $\bar{T}(b) \subset M(\bar{T}(a))$ is proved.

Changing rôles of a and b one shows similarly,

$a \subset N_d(b) \Rightarrow \bar{T}(a) \subset M(\bar{T}(b))$.

This completes the proof that $a \mapsto \bar{T}(a)$ is continuous.

ii) The continuity of a $\bar{T}(x,a)$ is proven with the same technique. (One even can set $d = e$.) □

(379) Lemma: Let $K \subset \bar{T}$ be compact. Then the mapping $PK \times PK \to PK^2$, $(A,B) \mapsto AB$ is continuous w.r. to the topologies induced by the canonical uniformities on PK resp. PK^2.

Proof: The multiplication $K \times K \to K^2$ is uniformly continuous ([BGT] II § 4.1 Th. 2). By [BGT] II § 2 Ex. 6d) the corresponding mapping $P(K \times K) \to PK^2$ is uniformly equicontinuous, hence its restriction to the subspace $PK \times PK$. □

Now we will consider minimal $(\bar{R},\subset,\bar{T})$-chains between x and y, $x \neq y$, of order $a \in \bar{R}_o$ and length n. By (374):
$\bar{T}(x,a)\ \bar{T}(a)^{n-2} \cap \bar{T}(y,a) = \emptyset$.

(3710) Lemma: $\overline{\bar{T}(x,a)\ \bar{T}(a)^{n-3}} \cap \overline{\bar{T}(y,a)} = \emptyset$.

Proof: In order to derive a contradiction we will assume, that some $\tau \in \bar{T}$ is contained in the above intersection.

From [BGT] II § 1.2, Prop. 2, Cor. 1 we infer
$\tau \in \overline{\bar{T}(x,a)\ \bar{T}(a)^{n-3}} \subset \bar{T}(x,a)\ \bar{T}(a)^{n-2}$. A product of open sets is open, thus $\bar{T}(x,a)\ \bar{T}(a)^{n-2}$ contains some τ-neighbourhood V. Because of $\tau \in \overline{\bar{T}(y,a)}$, V meets $\bar{T}(y,a)$ in contradiction to
$\bar{T}(x,a)\ \bar{T}(a)^{n-2} \cap \bar{T}(y,a) = \emptyset$. □

Now we can prove a type of continuity for $a \mapsto \lambda(x,y,a)$.

(3711) Proposition: Let be $x, y \in P$, $a \in \bar{R}_o$. Then there exists a $d \in \tilde{R}_o$ such that for all $b \in \bar{\bar{N}}_d(a) \cap \bar{R}_o$, $|\lambda(x,y,a)-\lambda(x,a,b)| \le 1$.

Proof: For each $A \subset \bar{T}$ the closure is given by
$\bar{A} = \bigcap_{N \in \bar{\mathcal{I}}^d} N(A)$ ([BGT] II § 1.2, Prop. 2, Cor. 1), hence (3710)
implies, that for some entourage $N \in \bar{\mathcal{I}}^d$,
$N(\bar{T}(x,a)\bar{T}(a)^{n-3}) \cap N(\bar{T}(y,a)) = \emptyset$.
For all $b \subset N_a(a)$ the subsets $\bar{T}(x,b)$ and $\bar{T}(b)$ are contained in some fixed compact $K \subset \bar{T}$ (this follows easily from [BGT] III § 4.5, Th. 1). Thus we may apply (377) and (379) in order to show that
$\bar{T}(x,b)\bar{T}(b)^{n-3} \subset N(\bar{T}(x,a)\bar{T}(a)^{n-3})$ and $\bar{T}(y,b) \subset N(\bar{T}(y,a))$ hold, if b is chosen "close to a", that is, if $b \subset N_d(a)$ and $a \subset N_d(b)$ for some appropriate $d \in \tilde{R}_o$, $d \subset a$.

Now the relation
$\bar{T}(x,b)\ \bar{T}(b)^{n-3} \cap \bar{T}(y,b) = \emptyset$
shows (374), that $\lambda(x,y,b) \geq n - 1$, hence $\lambda(x,y,b) \geq \lambda(x,y,a) - 1$. Since the conditions on a and b are symmetric, we infer likewise $\lambda(x,y,a) \geq \lambda(x,a,b) - 1$, which completes the proof. □

(3712) Proposition: Let R' be a dense subset of $\bar{R}_o$ and $x, y, u, v \in P$, $u \neq v$.
Then the net

$$\left(\frac{\lambda(x,y,a)}{\lambda(u,v,a)}\right)_{a \in \bar{R}_o} \quad \text{converges iff}$$

$$\left(\frac{\lambda(x,y,a)}{\lambda(u,v,a)}\right)_{a \in R'} \quad \text{converges.}$$

Proof: For the non-trivial implication choose for each $a \in \bar{R}_o$ some $b_a \in R'$ "near" the region a according to (3711). This yields a directed subset[R"]⊂[R']. The convergence of the $\bar{R}_o$-net now follows from

$$\frac{\lambda(x,y,b_a)-1}{\lambda(u,b,b_a)+1} \leq \frac{\lambda(x,y,a)}{\lambda(u,v,a)} \leq \frac{\lambda(x,y,b_a)+1}{\lambda(u,v,b_a)-1}$$

and $\lim_{b_a \in R'} \lambda(u,v,b_a) = \lim_{b_a \in R'} \lambda(x,y,b_a) = \infty$ (see (359)), if $x \neq y$. In the case $x = y$ clearly both nets converge to 0. □

Now from (3312) and (375) we can derive the main result of this section:

(3713) Theorem: The convergence of the chain quotient in the $(\tilde{R},\subset,\tilde{T})$-theory implies the convergence of the chain quotient in the $(\overline{R},\subset,\overline{T})$-theory. The limits coincide.

4. THE HELMHOLTZ-LIE PROBLEM

In section 3.6 we have developed the prerequisites to apply the Tits/Freudenthal solution of the Helmholtz-Lie problem. In order to study the relation between mobility and convergence of the chain quotient in section 4.2 it is convenient to give an exposition of the differential geometry of this problem. We shall follow [FRE] with two minor modifications: first we shall prove $\bar{T}$ to be a Lie group without using the classification of [FRE] and furthermore an argument concerning the dimension of orbits is reformulated using more recent results [MSZ].

4.1. IMPLICATIONS OF THE THEOREM OF YAMABE

The passage from a pure topological analysis of physical space to the differential geometrical view is made possible by the solution of Hilberts fifth problem, the topological characterization of Lie groups. After some decades of work by L. Brouwer, B. Kerekjarto, J. v. Neumann, L. Pontrjagin, C. Chevalley, D. Montgomery, L. Zippin et al. the most general solution was given by H. Yamabe [YAM].

In sequel we suppose $(R,<,T)$ to fulfil the axioms R1 to R5 (and R6 to be formulated) and use essentially only the propositions proved for $(\bar{R},\subset,\bar{T})$ in section 3.

A topological group (G,τ) will be called a <u>Lie group</u>, if G can be equipped with a differentiable structure δ such that (G,δ) is a Lie group (for definition see e.g. [B&C] 12.1) and if, in addition, the topology induced by δ coincides with τ. This applies especially to $(\bar{T},\bar{\mathcal{t}}) = (\bar{T},c)$.

(411) Theorem: There exists a basis $\mathcal{U}$ of the neighbourhood filter of Id $\in \bar{T}$ and an infinite net of subgroups

$(N_U)_{U \in \mathcal{U}}$, downwardly directed, with the following properties:

For each $U \in \mathcal{U}$,

(i) $N_U \subset U$,

(ii) N_U is a compact, connected, invariant subgroup,

(iii) $\bar{T}/N_U$ is a Lie group.

Further,

(iv) if $\bar{T}$ itself is not a Lie group, we may assume

$\forall\, U, V \in \mathcal{U}, U \subsetneqq V \Rightarrow N_U \subsetneqq N_V$, hence

$\forall\, U \in \mathcal{U}, N_U \neq \{Id\}$,

(v) if $\bar{T}$ itself is a Lie group we may (trivially) assume

$\forall\, U \in \mathcal{U}, \quad N_U = \{Id\}$.

The point (v) is only adduced for sake of uniformity of argument.

Proof: By [YAM] any locally compact group equipped with an Id-neighbourhood containing no non-trivial invariant subgroup is a Lie group. Let U be a compact Id-neighbourhood. Either we have the "trivial" case (v) or U contains an invariant subgroup $N_U \neq \{Id\}$. We may assume N_U is maximal in U, hence closed, hence compact. Let $\nu : \bar{T} \to \bar{T}/N_U$ be the (open) quotient map. The Id-neighbourhood $\nu[U]$ cannot contain any non-trivial invariant subgroup M, since in this case $\nu^{-1}(M)$ would be an invariant subgroup in U including the maximal N_U. Now [YAM] implies $\bar{T}/N_U$ is a Lie group, where the case $\nu[U] = \{Id\}$, i.e. $\bar{T}/N_U$ is discrete, is understood to be included.

If the assiquement $U \mapsto N_U$ is performed for any basis $\mathcal{U}'$, (iv) always can be achieved by passing to a coarser basis $\mathcal{U} \subset \mathcal{U}'$. If N_U is not connected it is to be replaced by its connected Id-component N_U^o. Because of the canonical isomorphism:

$$\frac{\bar{T}}{N_U} \cong \frac{\bar{T}/N_U^o}{N_U/N_U^o}$$

(see [BGT] III § 2.7, Prop. 2.2, Cor.), and the finiteness of N_U/N_U^o, since N_U is compact, the two topological groups $\bar{T}/N_U$ and $\bar{T}/N_U^o$ are locally isomorphic. Thus both $\bar{T}/N_U$ and also $\bar{T}/N_U^o$ are Lie groups. □

In sequel we consider some fixed $N = N_U$, $x \in P$ and $J = J_x$, the stability subgroup.

(412) Definition: We shall denote by P/N the space of orbits w.r. to the operation of the group N on P, endowed with the quotient topology.

(413) Proposition: There exist the following canonical homeomorphisms:

$$\frac{P}{N} \underset{(i)}{\cong} \frac{\bar{T}/J}{N} \underset{(ii)}{\cong} \frac{\bar{T}}{JN} \underset{(iii)}{\cong} \frac{\bar{T}/N}{JN/N} .$$

Further,

(iv) P/N is a Hausdorff space.

Proof:

(i) $P \cong \bar{T}/J$ w.r. to the map $\tau x \mapsto \tau J_x$, see (3613).

(iii) Since the subgroup JN operates continuously on $\bar{T}$ on the right, it follows by [BGT] III § 2.7. Prop. 22, that the canonical map

$\tau JN \mapsto \{\tau jN | j \in J\} = \{\tau NjN | jN \in JN/N\}, \quad \tau \in \bar{T},$

is a homeomorphism

$$\frac{\bar{T}}{JN} \cong \frac{\bar{T}/N}{JN/N} .$$

(ii) $\bar{T}$ operates on P/N by means of $Ny \mapsto \tau Ny = N\tau y$, since N is invariant. This operation is transitive, continuous by [BGT] III § 2.4 Prop. 11 and proper, which can easily be inferred from [BGT] III § 4.2 Prop. 5i). The "point" $Nx \in P/N$ has w.r. to this operation the stability subgroup JN = NJ.
Hence [BGT] III § 4.2 Prop. 4c) shows the canonical map
$\tau\ JN \mapsto N\tau x$ to be a homeomorphism
$\bar{T}/JN \cong P/N$ and by (i),
$\bar{T}/JN \cong (\bar{T}/J)/N$.

(iv) Being a product of two compact subsets, JN is a compact, hence closed subgroup . Therefore $\bar{T}/JN$ is a homogeneous Hausdorff space (see [BGT] III § 2.5 Prop. 13). □

The above proof of (iv) and (iii) shows that JN/N is a closed subgroup of $\bar{T}/N$. $P/N \cong \frac{\bar{T}/N}{JN/N}$ thus is a quotient of a Lie group and a closed subgroup and can be endowed with a C^∞-differential structure compatible with its topology (see [B&C] Prop. 12.9.4, including the discrete case). In brief:

(414) Corollary: P/N is a C^∞-manifold.

For later purpose we need the following general propositions.

(415) Definition: Let any group G operate on spaces X and Y. A mapping $f : X \to Y$ is called <u>equivariant</u> (re G) iff, for all $x \in X$ and $g \in G$, $f(g\cdot x) = g \cdot f(x)$ holds.

(416) Lemma: Let any group G operate on C^∞-manifolds M_1, M_2 by means of C^∞-diffeomorphisms $L_{i,g} : M_i \to M_i$, $i=1,2$, $g \in G$. Let $\mathcal{T} L_{i,g} : \mathcal{T} M_i \to \mathcal{T} M_i$ be the induced operation of G on

the tangent bundle and $f : M_1 \to M_2$ an equivariant C^∞-mapping. Then the derivation of f, $\mathcal{T}f : \mathcal{T}M_1 \to \mathcal{T}M_2$, is equivariant (w.r. to the operation of G on the tangent bundle).

Proof: The equivariance of f means $L_{2,g}^{-1} \circ f \circ L_{1,g} = f$. By the functorial property of $\mathcal{T}$ (see [B&C] 4.3 or [ABR] Th. 5.7) we have: $\mathcal{T}(L_{2,g})^{-1} \circ \mathcal{T}f \circ \mathcal{T}L_{1,g} = \mathcal{T}f$. □

(417) Corollary: Let $x_i \in M_i$ be fixed points of G and assume $f(x_1) = x_2$. Then $\mathcal{T}_{x_1} f : \mathcal{T}_{x_1} M_1 \to \mathcal{T}_{x_2} M_2$ is equivariant (w.r. to the induced operation of G on the tangent spaces $\mathcal{T}_{x_i} M_i$, $i=1,2$).

Now let us return to situation of (411) to (414), where $\mathcal{U}$ is the basis of Id-neighbourhoods of $\bar{T}$ according to (411). We shall write $e = \mathrm{Id} \in \bar{T}$.

(418) Proposition: Let $U, V \in \mathcal{U}$ such that $U \subset V$, hence $N_U \subset N_V$. Further consider the canonical quotient maps

$\nu_U : P \to P/N_U$ and

$\nu_V : P \to P/N_V$. Then there exists a canonical submersion (i.e. a surjection with maximal rank, see [B&C] 6.1)

$\nu_{U,V} : P/N_U \to P/N_V$ satisfying

(i) $\nu_{U,V} \circ \nu_U = \nu_V$,

(ii) $\nu_{U,V}$ is equivariant w.r. to the operation of $\bar{T}$,

(iii) let $\hat{y} \in P/N_V$, then $\nu_{U,V}^{-1}(\hat{y})$ is homeomorphic to JN_V/JN_U.

Proof: Consider the diagram

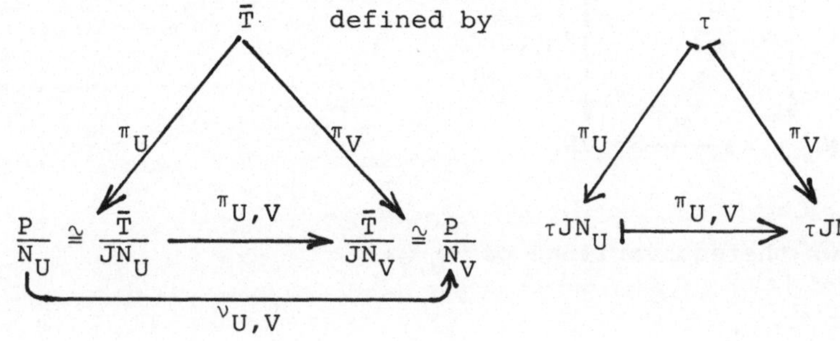

where (413) is used.

If the diagramm is commutative, $\pi_{U,V}$ is thereby uniquely determined. ν_U and ν_V are submersions. Considering for instance ν_U, this follows from the commutative diagram

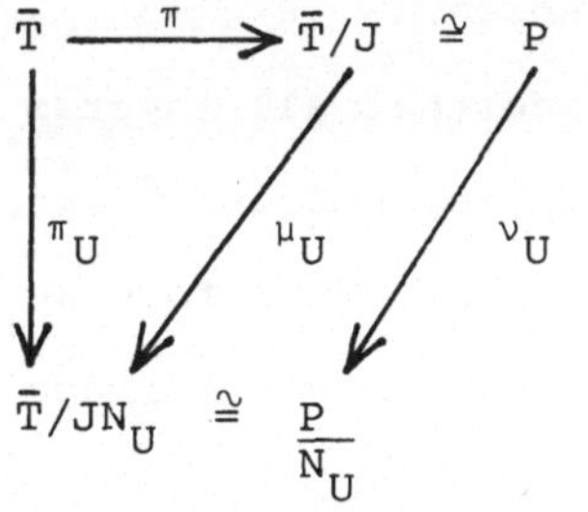

and the conclusion

"if π and π_U are submersions, then μ_U is a submersion" (see [B&C] Prop. 12.9.4, 6.1.2 and 6.1.3). The homeomorphisms denoted by $\cong$ clearly become C^∞-diffeomorphisms, hence ν_U is a submersion. Composition of diagrams now yields $\nu_{U,V} \circ \nu_U = \nu_V$, hence $\nu_{U,V}$ is also a submersion.

In order to prove the equivariance we use the formula for the C^∞-diffeomorphism $P/N \cong \bar{T}/JN$, namely $N\tau x \overset{\cong}{\longmapsto} \tau JN$ (see proof of (413)(ii)) and take $N = N_U$, resp. $N = N_V$ to get the diagram:

$$\begin{array}{ccc} N_U\,\tau\,x & \overset{\cong}{\longmapsto} & \tau\,JN_U \\ \Big\downarrow{\scriptstyle \nu_{U,V}} & & \Big\downarrow{\scriptstyle \nu_{U,V}} \\ N_V\,\tau\,x & \overset{\cong}{\longmapsto} & \tau\,JN_V \end{array}$$

This shows the equivariance of $\nu_{U,V}$:

$$\begin{array}{ccc} N_U y & \overset{L_\sigma}{\longmapsto} & N_U\,\sigma\,y \\ \Big\downarrow{\scriptstyle \nu_{U,V}} & & \Big\downarrow{\scriptstyle \nu_{U,V}} \\ N_V y & \overset{L_\sigma}{\longmapsto} & N_V\,\sigma\,y, \end{array}$$

where $y \in P$ and $\sigma \in \bar{T}$ is arbitrarily chosen.

In order to show property (iii) we notice that all inverse images $\nu_{U,V}^{-1}(\hat{y})$ of points $\hat{y} \in P/N_V$ are homeomorphic, because $\bar{T}$ operates transitively on P/N_V and $\nu_{U,V}$ is equivariant. Therefore we may take $\hat{y} = N_V x \overset{\cong}{\longmapsto} eJN_V$ in order to calculate the inverse image:

$$\nu_{U,V}^{-1}[y] \overset{\cong}{\longmapsto} \pi_{U,V}^{-1}[eJN_V]$$

$$= \pi_U[\pi_V^{-1}[eJN_V]]$$

$$= \pi_U[JN_V]$$

$$= JN_V/JN_U. \qquad \square$$

Now consider an arbitrarily chosen $U \in \mathcal{U}$. We shall simplify our notation by writing $N_U = N$, $Nx = \hat{x}$, $\bar{T}/N = \hat{T}$, $P/N = \hat{P}$, $J_x N/N = \hat{J}$.

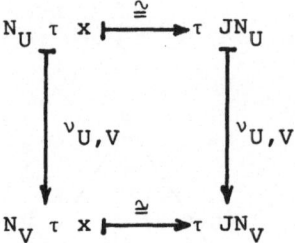

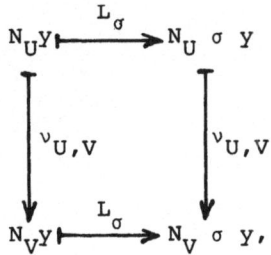

Let T be the Lie algebra of $\hat{T}$ and J the subalgebra belonging to the subgroup $\hat{J}$.

Consider any $\hat{j} \in \hat{J}$. The C^∞-mapping $\hat{t} \mapsto \hat{j}\,\hat{t}\,\hat{j}^{-1}$, $\hat{t} \in \hat{T}$, maps $\hat{e} \in \hat{T}$ onto $\hat{e}$, its derivation at the point $\hat{e}$ thus being a linear endomorphism of $T_{\hat{e}}\hat{T} \cong T$, which will be denoted by

$\tilde{j} : T \to T$

(419) Lemma: Let $t \in T$ and $\lambda \in \mathbb{R}$, then

$$\hat{j}\,\exp(\lambda t)\,\hat{j}^{-1} = \exp(\lambda\tilde{j}(t)).$$

Proof: The left hand side defines a 1-parameter-subgroup of $\hat{T}$, thus is of the form $\exp(\lambda t')$, $t' \in T$. $\tilde{j}(t) = t'$ follows by one of the possible definitions of the derivation $\tilde{j}$, namely by the induced transformation of smooth curves (see [ABR] Def. 5.6). □

The assignement $\hat{j} \mapsto \tilde{j}$ forms a linear representation of the group $\hat{J}$, which is a restriction of the adjoint representation of $\hat{T}$ (see [F&V] 9.6). It is faithful, since $\hat{J}$ contains no central elements except $\hat{e}$ ($\hat{j}\hat{\tau}\hat{x}=\hat{\tau}\hat{j}\hat{x}=\hat{\tau}\hat{x} \Rightarrow \hat{j}=\hat{e}$). Since $\hat{J}$ is compact, T may be equipped with a positively definite inner product, denoted by <,>, such that $\hat{J}$ is represented by <,>-orthogonal endomorphisms (see [F&V] 35.1). J is invariant w.r. to all $\tilde{j}$, hence this holds also for its <,>-orthogonal complement, which will be denoted by K.

Consider the following composed C^∞-map f:

$$K \hookrightarrow T \xrightarrow{\exp} \hat{T} \xrightarrow{\nu} \hat{T}/\hat{J} \cong \hat{P}, \text{ that is:}$$

(4110) Definition: $f(X) \underset{\text{def}}{=} (\exp X)\,\hat{x}$ for all $X \in K$.

Its derivation at 0, $f' \underset{\text{def}}{=} T_0 f$, decomposes in the following way:

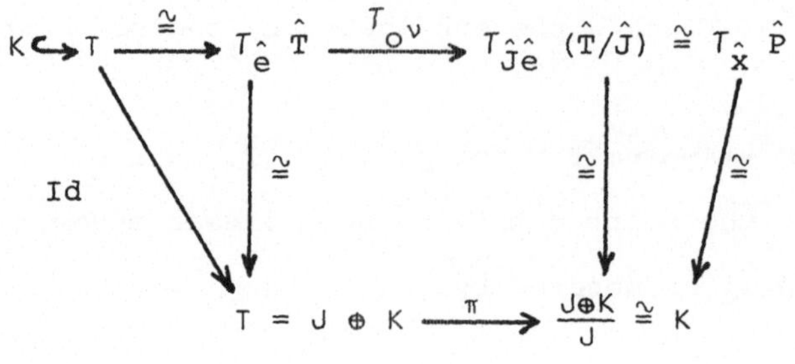

where π denotes the canonical quotient map $J \oplus K \to J \oplus K/J$. Therefore, up to isomorphisms, f' is equal to Id_K. Hence f is a local C^∞-diffeomorphism and for a suitable $\hat{x}$-neighbourhood U the restriction

$f^{-1}|U : U \to K \cong \mathbb{R}^p$

is a C^∞-chart of $\hat{P}$ around the point $\hat{x}$. It will be referred to as the <u>canonical</u> <u>chart</u>. It is even analytical, since exp and ν are locally analytic (see [K&N] 1, I 4.2.).

(4111) Proposition (compare [FRE] 2.7.):
$\hat{J}$ operates on $\hat{P}$ by locally orthogonal transformations via the canonical chart.

Proof: Since $\tilde{j}|K$ is orthogonal it suffices to show that $\hat{j}$ locally coincides with $\tilde{j}$ when computed in the canonical chart. This means:
$f^{-1}\ \hat{j}\ f(k) = \tilde{j}(k)$ for all $\hat{j} \in \hat{J}$ and $k \in K$.

It follows:

$$\begin{aligned}\hat{j}\ f(k) &\cong \hat{j}(\nu\circ\exp)(k)\\ &= \hat{j}(\exp k)\ \hat{J}\\ &= \hat{j}(\exp k)\ \hat{j}^{-1}\ \hat{J}.\end{aligned}$$

$$\begin{aligned}f^{-1}\ \hat{j}\ f(k) &= \log(\hat{j}(\exp k))\ \hat{j}^{-1}\\ &= \tilde{j}(k), \quad \text{by (419).} \quad \square\end{aligned}$$

(4112) Lemma: There exists a local C^∞-injection $\tau : U \to \hat{T}$ such that $\tau(\hat{y})(\hat{x}) = \hat{y}$ for all $\hat{y}$ in some neighbourhood of $\hat{x}$.

Proof: Being a submersion, ν has a local C^∞-section τ, that means: $\nu \circ \tau = \mathrm{Id}_U$ (see [B&C] Prop. 6.1.4.). Let $\hat{y} \in U$.
Identifying $\hat{T}/\hat{J} = \hat{P}$, we may express $\nu\ \tau(\hat{y}) = \hat{y}$ in the form $\tau(\hat{y})\hat{J} = \sigma\ \hat{J}$ where $\sigma \in \hat{T}$ such that $\sigma\hat{x} = \hat{y}$.
Hence $\tau(\hat{y}) = \sigma\hat{j}, \hat{j} \in \hat{J}$ and $\tau(\hat{y})(\hat{x}) = \sigma\hat{j}(\hat{x}) = \sigma(\hat{x}) = \hat{y}$. □

(4113) Theorem: $\hat{P}$ may be equipped with a Riemannian metric g, such that $\hat{T}$ operates isometrically on $\hat{P}$.

Proof (see [F&V] 6.4.3.1, [K&N] 2, X 3.): The positive definite inner product <,> on $K \cong T_{\hat{x}}\hat{P}$ has to be transferred to the other tangent spaces $T_{\hat{z}}\hat{P}$, $\hat{z} \in \hat{P}$. This may be done by the linear isomorphism $T_{\hat{x}} f : T_{\hat{x}}\hat{P} \to T_{\hat{z}}\hat{P}$, where $f \in \hat{T}$, $f\hat{x} = \hat{z}$.
The construction does not depend on the choice of f. Let $g \in \hat{T}$ also satisfying $g\hat{x} = \hat{z}$, then $f^{-1} g\hat{x} = \hat{x}$, $j \underset{\text{def}}{=} f^{1}g \in \hat{J}$,
$T_{\hat{x}}\, g = T_{\hat{x}}\, f \circ T_{\hat{x}}\, j$, where $T_{\hat{x}}\, j$ leaves the inner product in $T_{\hat{x}}\hat{P}$ invariant by (4110).
In this way an inner product $g_{\hat{z}}$ on $T_{\hat{z}}\hat{P}$ is defined globally. The transporting maps can be chosen to depend locally on $\hat{z}$ in a C^∞-differentiable way by use of (4112), hence $\hat{z} \mapsto g_{\hat{z}}$ is C^∞ too.
Consider any $\hat{z} \in \hat{P}$ and $h \in \hat{T}$. Let $g \in \hat{T}$ such that $g\hat{x} = \hat{z}$. Together with $(T_{\hat{x}}g)^{-1} = T_{\hat{z}}\, g^{-1}$ and $T_{\hat{x}}(h \circ g)$, the product $T_{\hat{x}}(h \circ g) \circ T_{\hat{z}}\, g^{-1} = T_{\hat{z}}\, h$ is an isometry.
Hence $\hat{T}$ operates isometrically on $(\hat{P}, g)$. □

Whereas g is uniquely determined by the inner product <,> on $T_{\hat{x}}P$ and the claim that $\hat{T}$ consists of isometries, the inner product <,> is only fixed up to positive definite linear maps commuting with the representation of $\hat{J}$ on K. At any case, a positive scaling factor remains facultative.

4.2. MOBILITY AND DISTANCE MEASURED BY CHAINS

We now turn to the last and decisive axiom of Freudenthal. It is developed from earlier postulates claiming for instance that two sets of points, whose internal distances are pairwise equal, may be mapped onto each other by a (global) isometry (Birkhoff). Without using the notion of a metric, "mobility" is now formulated by Freudenthal (and similarly by Tits) in the following way:

(421) There exist two points $x, y \in P$ such that the orbit $J_x y$ dissects the space P,

where "dissection" means:

(422) Definition: Let B be a topological space, $A \subset B$. A is said to dissect B, iff $B \setminus A$ is not connected.

The postulate (421) is intimately connected with the possibility of measuring distances. The convergence of the chain quotient (see (3514)) cannot be proved on the basis of the axioms R1 to R5, as is shown by the following

(423) Counter example: R is the lattice of bounded open subsets of $\mathbb{R}^2$, T is induced by translations.

Since the chain links do not freely rotate, the limit of

$$\left(\frac{\lambda(x,y,a)}{\lambda(u,v,a)}\right)_{a \in R_o}$$

depends on the shape of the links as is shown in the following figure:

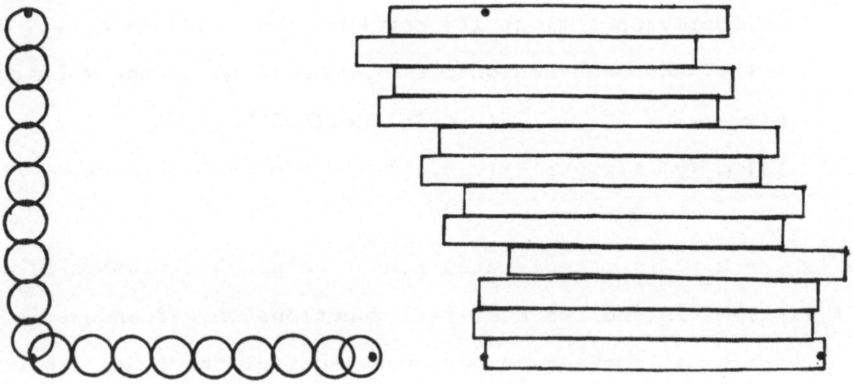

fig. (424)

Clearly, the counter example does not satisfy the requirement (421). This is of course not accidental as convergence of the chain quotient and mobility are essentially equivalent, as we will show.

(425) Theorem: In a system in which axioms R1 to R5 hold, the following 3 assertions are equivalent:

(i) There exist two points $x, y \in P$ such that the orbit $J_x y$ dissects the space P (Freudenthal).

(ii) There exists a point $x \in P$ and a neighbourhood V of x such that each orbit $J_x y$, $x \neq y \in V$, dissects the space P (Tits). Further, $\bar{T}$ is a Lie group.

(iii) The chain quotient w.r. to $(R,<,T)$ converges uniformly on regions (in the sense defined below), or P is homeomorphic to the real line.

Explicitly, the meaning of (iii) is contained in the following

(426) Definition: The $(R,<,T)$-chain quotient is said to converge uniformly on regions iff for some $x \in P$ and each $u, v \in P$, $u \neq v$, and each region $f \in \tilde{R}_o$ containing x, the mapping $Q(x,-,-) : [R_o] \to \mathcal{F}_c(P \setminus f, \mathbb{R})$ defined by $[a] \mapsto (y \mapsto \lambda(x,y,a)/\lambda(u,v,a))$ is Cauchy-convergent.

Here $[R_o] = R_o/T$ is the directed set of congruence classes of regions and $\mathcal{F}_c(P \setminus f, \mathbb{R})$ is the space of real functions on $P \setminus f$ endowed with the uniformity of compact convergence (equivalently: uniform convergence on regions).

The Cauchy-convergence of the mapping $Q(x,-,-)$ will be understood in the sense of [BGT] I § 7.3, that is, the image under $Q(x,-,-)$ of the section filter on $[R_o]$ is a Cauchy filter base on $\mathcal{F}_c(P \setminus f, \mathbb{R})$.
The detailed formulation of (426) thus reads as follows:

(427) $\exists\, x \in P \;\forall\, u, v \in P$ such that $u \neq v \;\; \forall\, f \in \tilde{R}_o$ such that $x \in f \;\; \forall\, b \in \tilde{R}_o$ such that $b \cap f = \emptyset \;\; \forall\, \varepsilon > 0$ $\exists\, a_o \in R_o \;\forall\, a_1, a_2 \in R_o$ such that $a_1 < a_o$ and $a_2 < a_o \;\forall\, y \in b$:

$$\left| \frac{\lambda(x,y,a_1)}{\lambda(u,v,a_1)} - \frac{\lambda(x,a,a_2)}{\lambda(u,v,a_2)} \right| < \varepsilon \quad .$$

By [BGT] X § 1. Th. 1, and the completeness of $\mathbb{R}$, the space $\mathcal{F}_c(P \setminus f, \mathbb{R})$ is complete and we may conclude the following

(428) Proposition: If (427) is fulfilled, the chain quotient converges to a function $\Lambda(x,y,u,v)$.

(429) Theorem: Under the assumptions of (428), $\Lambda(x,-,u,v) \in \mathcal{C}_c(P \setminus f, \mathbb{R})$, the space of continuous functions on $P \setminus f$.

Proof: $P\setminus f$ is locally compact and by [BGT] X, § 1.6, Th. 2, Cor. 3, $C(P\setminus f,\mathbb{R})$ is closed in $F_c(P\setminus f,\mathbb{R})$. It suffices therefore to approximate $\lambda(x,y,a)/\lambda(u,v,a)$ by a continuous function such that the error converges to 0 w.r. to the section filter on $[R_0]$.

Consider the 2 closed sets $\overline{N_a x}$ and $\complement N_a^2 x$. P being paracompact (3614), hence normal ([BGT] IX § 4.4, Prop. 4), there exists a continuous function $\varphi : P \to [1,3]$ which is equal to 1 at every point of $\overline{N_a x}$ and equal to 3 at every point of $\complement N_a^2 x$. Hence $\sup\{|\lambda(x,y,a)-\varphi(y)|\,|y\in N_a^3 x\} \le 1$.

In this way, we inductively construct a continuous function $\tilde{\lambda} : P \to \mathbb{R}$ with the property $\sup\{|\lambda(x,y,a)-\tilde{\lambda}(y)|\,|y\in P\} \le 1$.

Hence $\sup\left\{\left|\frac{\lambda(x,y,a)}{\lambda(u,v,a)} - \frac{\tilde{\lambda}(y)}{\lambda(u,v,a)}\right| \,\middle|\, y\in P\setminus f\right\} \le \frac{1}{\lambda(u,v,a)}$

and the approximation of the chain quotient by a continuous function is arbitrarily close. The approximation is even performed w.r. to the uniformity of uniform convergence. □

It is not known whether (425)(iii) can be weakened to pointwise convergence of the chain quotient. Note that convergence is not given in the case of one-dimensional P, that is $P \cong S^1$ or $P \cong \mathbb{R}$ (see (4214)). This is due to the existence of infinite fractions: in the case $P \cong \mathbb{R}$ for instance one can easily compute

$$\lim_{n\to\infty} \frac{\lambda(0,\frac{2}{3},a_n)}{\lambda(0,1,a_n)} = \infty, \text{ where}$$

$$a_n \underset{\text{def}}{=} \;]-n^{-2}2^{-n}, n^{-2}2^{-n}[\; \cup \;]2^{-n}(1-n^{-2}), 2^{-n}(1+n^{-2})[$$

and the representation $\frac{2}{3} = 0.101010\ldots$ in binary digits is used.

This is the reason for including the case $P \cong \mathbb{R}$ in (425)(iii), whereas the case $P \cong S^1$ is ruled out by (425)(i).

The remainder of section 4.2 is dedicated to the proof of theorem (425).

4.2.1. PROOF OF "(i) ⇒ (ii)"

We shall use the notations $\hat{T}$, $\hat{x} \in \hat{P}$, $\hat{J}$ defined in section 4.1.

(4211) Proposition: There exists an Id-neighbourhood $U_o \in \mathcal{U}$ ($\mathcal{U}$ is the basis according to (411)) such that for each $U \in \mathcal{U}$, $U \subset U_o$, there exists a $y \in P$ such that the corresponding orbit $\hat{J}\,\hat{y}$ dissects $\hat{P}$.

Without loss of generality we may assume $\forall\, U \in \mathcal{U}$, $U \subset U_o$ and we will use this convention henceforth.

Proof: According to (425)(i) some orbit $J_x y = J\,y$ dissect P. Since $J\,y$ is closed, $P \setminus J\,y$ is an open, non-connected subset.
This means ([BGT] I § 11.1) that it is of the form $P \setminus J\,y = A \cup B$, A and B being open, non-empty, disjoint subsets. The quotient map $\nu : P \to \hat{P} = P \setminus N$ is surjective and open. Hence
$\hat{P} \setminus \hat{J}\,\hat{y} = \hat{P} \setminus \nu[Jy] \subset \nu[P \setminus J_y] = \nu[A] \cup \nu[B]$, where $\nu[A]$ and $\nu[B]$ are open subsets of $\hat{P}$. In order to insure the non-connectedness of $\hat{P} \setminus \nu[Jy]$ if suffices to show (see [BGT] I § 11.1 Def. 2) that $(\hat{P} \setminus \nu[Jy] \cap \nu[A]$ and $(\hat{P} \setminus \nu[Jy]) \cap \nu[B]$ are non-empty and $\nu[A] \cap \nu[B] \cap (\hat{P} \setminus \nu[Jy]) = \emptyset$.
In order to show the latter, assume
$\hat{z} = Nz \in \nu[A] \cap \nu[B] \cap (\hat{P} \setminus \nu[Jy])$. It follows that Nz meets A and B but not Jy, hence $Nz \subset P \setminus Jy = A \uplus B$ and Nz is non-connected in contradiction to (411)(ii).
In order to show $(\hat{P} \setminus \nu[Jy]) \cap \nu[A] \neq \emptyset$ assume $\nu[A] \subset \nu[Jy]$. This means that any $a \in A$ lies on some orbit $N\,j\,y$, $j \in J$. But there

exists a neighbourhood V of a, disjoint from Jy and Na $\subset$ V if N is chosen sufficiently small. This is a contradiction. □

We will use the notion of dimension of a set A (in symbols: dim A) in the sense of the Hurewicz and Wallmann [H&W], where "dimension" is inductively defined by the dimension of the boundary of A. The restriction in [H&W] to separable metric spaces is not too limited for our purposes, since $\hat{P}$ is a connected, paracompact Riemannian manifold (see [B&C] 3.4).
For topological manifolds the notion of dimension in [H&W] coincides with the usual notion of dimension of the manifold ([H&W] Th. IV 3. Cor. 1).
From [H&W] Th. IV 4. Cor. 1 we infer immediately:

(4212) Proposition: Let be $\dim \hat{P} \underset{def}{=} p$ and $\hat{J}\hat{y}$ dissecting $\hat{P}$, then $\dim \hat{J}\hat{y} \geq p - 1$ holds.

We can now apply the results on orbits of compact transformation groups in [MSZ]. Let us assume that for some neighbourhood V of $\hat{x}$, $\dim \hat{J}\hat{y} \leq p - 2$ for all $\hat{y} \in V$. This means that the set $\{\hat{y} \in \hat{P} \mid \dim \hat{J}\hat{y} \leq p-2\}$ contains an interior point and by [MSZ], Th. 1 Cor., each orbit of $\hat{J}$ has dimension at most $p - 2$, in contradiction to (4211) and (4212). This proves the following

(4213) Lemma: Each neighbourhood of $\hat{x}$ contains some point $\hat{y}$ such that $\dim \hat{J}\hat{y} \geq p - 1$ holds.

We shall discuss the case $p = 1$ separately. The following holds without assuming (425)(i).

(4214) Proposition: If $\dim \hat{P} = 1$, $\hat{P}$ is isometrically isomorphic either to the real line $\mathbb{R}$ or the circle S^1.

Proof: The group of isometries of a connected, p-dimensional Riemannian manifold is of dimension at most $\frac{1}{2}p(p+1)$ (see [K&N] VI. Th. 3.3, 3.4).

In the case $p = 1$ it follows that $\dim \hat{T} = 1$ because $\hat{T}$ operates transitively on $\hat{P}$. Thus the maximal dimension of $\hat{T}$ is attained and in this case [K&N], Note 10 Th. 1, states that $\hat{P}$ is isometric either to $\mathbb{R}^1$ or S^1, since the hyperbolic and elliptic spaces of dimension 1 coincide with these. □

(4215) Corollary: If $\bar{T}$ is a Lie group and $\dim P = 1$, it follows that $P \cong \mathbb{R}$ and hence (425)(ii) is fulfilled.

Now we consider the case $p > 1$.

We shall identify K and $\mathbb{R}^p$ (see (4110)) such that the inner product $<,>$ on K coincides with the usual inner product on $\mathbb{R}^p$. We recall that w.r. to the canonical chart $f^{-1} : U \to \mathbb{R}^p$, $f^{-1}\hat{J}f$ consists of linear, orthogonal, locally defined mappings $\mathbb{R}^p \to \mathbb{R}^p$. Let be $\hat{y} \in U$ and $\dim \hat{J}\hat{y} \geq p - 1$ according to (4213). We define $r = \|f^{-1}\hat{y}\|$ and $S(0,r) \underset{def}{=} \{\xi \in \mathbb{R}^p \mid \|\xi\| = r\}$. We may choose $\hat{y} \in \hat{P}$ such that $f^{-1}[U]$ contains $S(0,r)$ and hence the image of the orbit $Y \underset{def}{=} f^{-1}\hat{J}\hat{y} \subset S(0,r)$. Y is a compact submanifold of $S(0,r)$ of dimension $p - 1$, hence open in $S(0,r)$. By virtue of $p > 1$, $S(0,r)$ is connected and we infer $Y = S(0,r)$. The contractions $\xi \in \mathbb{R}^p \mapsto \lambda\xi$, $0 < \lambda < 1$ map $\hat{J}$-orbits onto $\hat{J}$-orbit, since $\hat{J}$ operates linearly. Hence in some neighbourhood $f^{-1}[V] \subset f^{-1}[U]$ each orbit is a sphere w.r. to the canonical chart and thus dissects $\hat{P}$. This proves:

(4216) Proposition: In a certain neighbourhood V of $\hat{x}$ each orbit $\hat{J}\hat{y}$, $\hat{x} \neq \hat{y} \in V$, dissects $\hat{P}$, if $\dim \hat{P} > 1$.

Now we are ready if $\bar{T}$ can be shown to be a Lie group, since then $\bar{T} = \hat{T}$ and (425)(ii) follow either by (4215) for $p = 1$ or by (4216) for $p > 1$.

(4217) Lemma: For all $U, V \in \mathcal{U}$ the following holds: $\dim P/N_V = 0$ or $\dim P/N_V = \dim P/N_U$.

Proof: Let us assume $U \subset V$ and $p = \dim P/N_U > \dim P/N_V = q > 0$.

The submersion $\nu_{U,V} : P/N_U \to P/N_V$ is equivariant w.r. to the operation of J (see (418)(ii)) and hence it maps J-orbits onto J-orbits. By (417) the same holds for its derivation $T_{N_U x}\, \nu_{U,V} : T_{N_U x}(P/N_U) \to T_{N_V x}(P/N_V)$, which can be identified with a non-trivial linear projection $\pi : \mathbb{R}^p \to \mathbb{R}^q$. As mentioned before, the J-orbits are the spheres in $\mathbb{R}^p$ resp. $\mathbb{R}^q$. But a sphere S^{p-1} is not projectable onto a sphere S^{q-1}, $0 < q < p$, because for instance $\pi^{-1}(0) \cap S^{p-1} \neq \emptyset$ implies $0 \in \pi[S^{p-1}] = S^{q-1}$, which is a contradiction. □

(4218) Theorem: $\bar{T}$ is a Lie group.

Proof: Let us assume the contrary and consider a countable infinite directed sub-family of $\mathcal{U}$ according to (411) and call the corresponding invariant subgroups N_i, $i \in \mathbb{N}$. By (4217) the sub-family may be chosen in such a way that the spaces P/N_i have the same dimension. Hence the submersions $\nu_{i+1,i} : P/N_{i+1} \to P/N_i$ have discrete inverse images (see [B&C] Prop. 6.2.1.).

By (418)(ii), JN_i/JN_{i+1} is discrete. Further, JN_{i+1} is a compact and open (see [BGT] III § 2.5 Prop. 14) subgroup of JN_i. But JN_i as a compact space is the topological sum of only a finite number of spaces and consequently contains only a finite number of different subgroups of the form JN_j. We conclude that $JN_{i+1} \subsetneqq JN_i$ is valid only for finitely many $i \in \mathbb{N}$. Thus we may assume $JN_k = JN_{k+1}$ for all $k \geq k_o$.

By (413) we have $P/N \cong \bar{T}/JN$, hence $P/N_k = P/N_{k+1}$, which means that for all $k \geq K_o$ and $y \in P$, $N_k\, y = N_{k+1}\, y$. Because $N_k \neq \{Id\}$, there exist two different points $y, z \in N_k\, y$ and a neighbourhood W of y such that $z \notin W$. Let V be a kernel of W (see (313)) containing z. If we construct the subfamily of $\mathcal{U}$ in such a way, that $N_k \subset \bar{T}(V)$ for some $k \geq k_o$, we obtain a contradiction, because $N_k\, z$ is contained in W and $y \in N_k\, z$ is impossible. □

4.2.2. PROOF OF "(ii) ⇒ (iii)"

By (4215) we may confine ourselves to the case dim $P > 1$. We recall that P is a complete, connected Riemannian manifold on which $\bar{T}$ operates isometrically. The corresponding metric shall be denoted by $\delta : P \times P \to \mathbb{R}$.

We consider two points $x, y \in P$ and a region $a_o \in R_o$ which will later be assumed to be sufficiently small (depending on ε). Further let $a \in R_o$ with $a < a_o$, be arbitrarily chosen. By (3713) we may consider $(\tilde{R}, \subset, \bar{T})$-chains instead of $(R, <, T)$-chains. We shall write a for $\tilde{a}$ etc.

One defines $\Delta(a) = \sup\{\delta(r,s) \mid r,s \in a\}$ as the <u>diameter</u> of the region a. It is finite since a is relatively compact. Let $[x,y,a,\tau_1 \ldots \tau_n]$ be a minimal chain. The triangle inequality implies the following

(4221) Lemma: $\delta(x,y) \leq \lambda(x,y,a) \cdot \Delta(a)$

In order to find an upper bound for $\lambda(x,y,a)$ we have to construct a chain between x and y which is "almost" minimal. This will be done by using a geodesic joining x and y. By [K&N] 1, IV Th. 4.2. there exists a geodesic $\gamma : [0,d] \to P$ such that $\gamma(0) = x$, $\gamma(d) = y$ and $d = \delta(x,y)$. Let V be the neighbourhood of x, in which the J_x-orbits dissecting P (see (4216)) are spheres $S(0,r)$, $0 < r < r_o$, w.r. to the canonical chart f^{-1}.

J_x consists of δ-isometries, hence all points of the orbit $f[S(0,r)]$ have the same distance $\delta(r)$ to x. $r \mapsto \delta(r)$ is a continuous map, because f is continuous. We choose $r_o > 0$ so small, that each geodesic of length $l \leq \delta(r_o)$ is an isometric map ("minimizing" in [K&N]). This is possible because there always exists "convex" neighbourhoods ([K&N] 1, IV Th. 3.6) and P is a homogeneous space.

(4222) Lemma: $r_1 < r_2 \Leftrightarrow \delta(r_1) < \delta(r_2)$.

Proof: Assume $r_1 < r_2$, $z_2 \in f[S(0,r_2)]$ and $\alpha : [0,\delta(r_2)] \to P$ being a geodesic such that $\alpha(0) = x$, $\alpha(\delta(r_2)) = z_2$. It meets the orbit $f[S(0,r_1)]$ at a point $z_1 \neq z_2$. Hence
$\delta(r_1) = \delta(x,z_1) < \delta(x,z_2) = \delta(r_2)$.
The converse holds for any strictly monotone real function. □

Thus locally the following notions are synonymous:
Orbit $J_x\, y$,
Image w.r. to f of the sphere $S(0,r) \subset \mathbb{R}^p$,
Equidistant sphere $\{z \in P \mid \delta(x,z) = \delta(r)\}$.

Now assume $a_o \in R_o$ is sufficiently small that $\Delta(a) \leq \Delta(a_o) < \frac{1}{3}\,\delta(r_o)$. Further let $x_o, y_o \in \bar{a}$ be points of maximal distance $\delta(x_o,y_o) = \Delta(a)$. These exist because $\bar{a}$ is compact. Consider a congruent mapping

$\tau \in \bar{T}$ such that $\tau x_o = x$. If $y_1 \underset{def}{=} \tau y_o$, $\delta(x,y_1) = \delta(x_o,y_o) < \delta(r_o)$ implies $J_x\, y_1 = f[S(0,r)]$, $0 < r < r_o$.

Define $K(0,r) = \{\xi \in \mathbb{R}^p \mid \|\xi\| < r\}$.

Either the geodesic $\gamma[0,d]$ is contained in $f[K(0,r_o)]$ or it meets $f[S(0,r)]$ at a point x_1, since $f[S(0,r)] = J_x\, y_1$ dissects P.

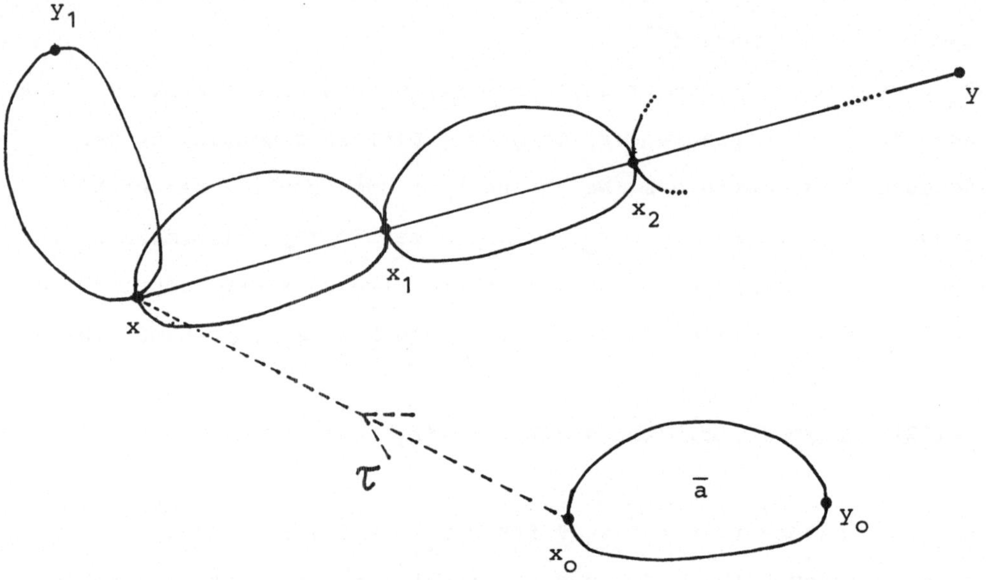

fig. (4223)

In the first case we put $m = 0$ and proceed as indicated below.

In the second case we have $x_1 = \gamma(t)$, $0 < t \le d$. Because $x_1 \in J_x\, y_1$ there exists some $j \in J_x$ such that $x_1 = j\, y_1$. We put $\tau_1 = j\tau$ for the chain to be constructed. If $y = x_1$, we are finished.

Otherwise we repeat the construction substituting x_1 for x, $\gamma|[\Delta(a),d]$ for γ etc.

Evidently, after m steps we obtain a chain between x and x_m of order $\bar{a}$ such that $\delta(x_m,y) < \Delta(a)$ and

(4224) $\quad m = \left[\frac{\delta(x,y)}{\Delta(a)}\right]$.

The latter follows, because $x, x_1, \ldots x_m, y$ are points on an (isometric) geodesic and $\delta(x,x_1) = \delta(x_1,x_2) = \ldots \delta(x_{m-1},x_m) = \delta(a)$, hence $\delta(x,y) = \delta(x,x_1) + \delta(x_1,x_2) + \ldots \delta(x_{m-1},x_m) + \delta(x_{m,y})$.

Now we have to construct the last two congruent mappings of the chain, τ_{m+1}, τ_{m+2}. Let be φ, $\psi \in \bar{T}$

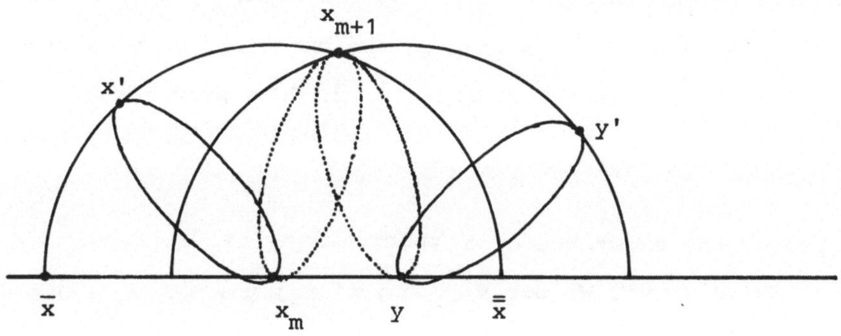

fig. (4225)

such that $\varphi\, x_o = x_m$, $\psi\, x_o = y$ and put $x' = \varphi\, y_o$, $y' = \psi\, y_o$. We need the following

(4226) Lemma: $\quad J_{x_m}\, x' \cap J_y\, y' \neq \emptyset$.

Proof: The geodesic $\gamma|[\delta(x,x_m),d]$ can be extended to a geodesic defined for all $t \in \mathbb{R}$, since P is complete ([K&N] 1, IV Th. 4.1.). Consider extensions at both ends x_m, y by a length of $\Delta(a)$, thus obtaining a geodesic

$\gamma' : [\delta(x,x_m)-\Delta(a), d+\Delta(a)] \to \mathbb{R}$ with length $2\Delta(a) + \delta(x_m,y) < 3\Delta(a) < \delta(r_o)$. By the choice of r_o, γ' is isometric. Hence the points

$\bar{x} \underset{def}{=} \gamma'(\delta(x,x_m)-\Delta(a))$ and

$\bar{\bar{x}} \underset{def}{=} \gamma'(\delta(x,x_m)+\Delta(a))$ satisfy

$\delta(x_m,\bar{x}) = \delta(x_m,\bar{\bar{x}}) = \Delta(a)$ and therefore lie on the orbit $J_{x_m} x'$.

Further:

$\delta(\bar{x},y) = \delta(\bar{x},x_m) + \delta(x_m,y) > \Delta(a)$ and

$\delta(\bar{\bar{x}},y) = \delta(\bar{\bar{x}},x_m) - \delta(x_m,y) < \Delta(a)$.

Hence $\bar{\bar{x}}$ lies in the interior of the sphere around y with radius $\Delta(a)$, which is just $J_y y'$, $\bar{x}$ lies in the exterior, $J_y y'$ dissects P and $J_{x_m} x'$ is connected and contains $\bar{x}$, $\bar{\bar{x}}$. This proves the assertion. □

(4226) proves the existence of a point $x_{m+1} \in J_{x_m} x' \cap J_y y'$. We put $x_{m+1} = \bar{j} x' = \bar{\bar{j}} y'$, $\bar{j} \in J_{x_m}$, $\bar{\bar{j}} \in J_y$, and $\tau_{m+1} = \bar{j} \varphi$, $\tau_{m+2} = \bar{\bar{j}} \psi$. This yields the chain $[x,y,a,\tau_1,\ldots\tau_{\bar{m}}]$ where

$$\bar{m} = \begin{cases} m & \text{if } x_m = y, \\ m+2 & \text{if } x_m \neq y. \end{cases}$$

A minimal chain between x and y must be shorter (not strictly), hence $\lambda(x,y,\bar{a}) \le \bar{m}$, and, by (4224):

(4227) $$\lambda(x,y,\bar{a}) \le \left[\frac{\delta(x,y)}{\Delta(a)}\right] + 2.$$

Let $d \in R_o$ be such that $a' = N_d(a)$ satisfies $|\lambda(x,y,a)-\lambda(x,y,a')| \le 1$ (see (3711)). $\bar{a} \subset N_d(a)$ implies $\lambda(x,y,\bar{a}) \ge \lambda(x,y,a')$, hence

$$\lambda(x,y,a) \le \left[\frac{\delta(x,y)}{\Delta(a)}\right] + 3 \le \frac{\delta(x,y)}{\Delta(a)} + 3.$$

Together with (4221) follows:

$$(4228)\quad \begin{cases} \dfrac{\delta(x,y)}{\Delta(a)} \le \lambda(x,y,a) \le \dfrac{\delta(x,y)}{\Delta(a)} + 3 \quad \text{and} \\ \dfrac{\delta(u,v)}{\Delta(a)} \le \lambda(u,v,a) \le \dfrac{\delta(u,v)}{\Delta(a)} + 3, \end{cases}$$

if u, v $\in$ P is another pair of points.

We will assume $x \neq y$ and $u \neq v$ in sequel. Let us write:

$\lambda(x,y,a) = \lambda_1$

$\lambda(u,v,a) = \lambda_2$

$\Delta(a) = \Delta$

$\delta(x,y) = \delta_1$

$\delta(u,v) = \delta_2$. We obtain:

$$A \underset{\text{def}}{=} \frac{\Delta}{(\delta_2+3\Delta)} \frac{\delta_1}{} \le \frac{\lambda_1}{\lambda_2} \le \frac{(\delta_1+3\Delta)}{\Delta} \frac{\Delta}{\delta_2} \underset{\text{def}}{=} B,$$

$$A = \frac{\delta_1}{\delta_2}\left(\frac{1}{1+\frac{3\Delta}{\delta_2}}\right) \ge \frac{\delta_1}{\delta_2} \frac{1}{1+\frac{3}{\lambda_2-3}} = \frac{\delta_1}{\delta_2}\left(1-\frac{3}{\lambda_2}\right),$$

From $\lambda_1 - 3 \le \frac{\delta_1}{\Delta}$ follows $\Delta \le \frac{\delta_1}{\lambda_1-3}$, hence:

$$B \le \frac{\delta_1 + \frac{3\delta_1}{\lambda_1-3}}{\delta_2} = \frac{\delta_1}{\delta_2}\left(1+\frac{3}{\lambda_1-3}\right).$$

This proves:

$$(4229)\qquad -\frac{\delta_1}{\delta_2}\frac{3}{\lambda_2} \le \frac{\lambda_1}{\lambda_2} - \frac{\delta_1}{\delta_2} \le \frac{\delta_1}{\delta_2}\frac{3}{\lambda_1-3}$$

By proposition (359), λ_1 and λ_2 increase indefinitely, as $a \in R_o$ decreases and we obtain:

(42210) $\lim_{a\in R_o} \frac{\lambda(x,y,a)}{\lambda(u,v,a)} = \frac{\delta(x,y)}{\delta(u,v)}$.

In the case x = y, which was hitherto excluded, $\lambda(x,y,a)$ is (constant and) equal to 1, whereas $\lambda(u,v,a)$ tends toward infinity. Hence the limit (42210) exists and is equal to zero.

In order to prove the uniform convergence on regions in (425)(iii), assume $b, c \in \bar{R}$, $\bar{b} \cap \bar{c} = \emptyset$, $x \in c$, $y \in b$. Now $\{\delta(x,y) \mid y \in b\}$ is bounded from above by, say, $\delta > 0$.

Let $\varepsilon > 0$ be given and chose some $K \in \mathbb{N}$, $K \geq \frac{3\delta}{\delta(u,v)} + 3$, and $a_o \in R_o$ so small that

$\lambda(u,v,a_o) \geq \frac{3\delta}{\delta(u,v)}$ and $\forall\, y \in b$, $\lambda(x,y,a_o) \geq \frac{3\delta}{\delta(u,v)} + 3$.

The latter for instance holds, if in addition $N^K_{a_o}(x) \subset c$. For any $a \in R_o$, $a \subset a_o$, it follows by (4229) that:

$$\left|\frac{\lambda_1}{\lambda_2} - \frac{\delta_1}{\delta_2}\right| \leq \frac{\delta_1}{\delta_2} \min\left\{\frac{3}{\lambda_2}, \frac{3}{\lambda_1 - 3}\right\} \leq \min\left\{\frac{3\delta}{\lambda_2\delta_2}, \frac{3\delta}{(\lambda_1-3)\delta_2}\right\} \leq \min\{\varepsilon,\varepsilon\} = \varepsilon .$$

4.2.3. PROOF OF "(iii) ⇒ (i)"

A homeomorphism $P \cong \mathbb{R}$ can be replaced by an isometry (see (4214)) and in this case (425)(i) clearly holds. Hence we may confine ourselves to the case where the chain quotient converges.

According to (3514), (3713) and (428) we will write

$$\lim_{a\in \bar{R}_o} \frac{\lambda(x,y,a)}{\lambda(u,v,a)} = \Lambda(x,y,u,v) = d(x,y).$$

The pair $u,v \in P$, $u \neq v$, is kept fixed in sequel, if not mentioned otherwise.

(4231) Theorem:

(i) $d : P \times P \to \mathbb{R}$ is a metric,

(ii) $\bar{T}$ operates d-isometrically on P,

(iii) if another pair u', v' ∈ P is chosen, the corresponding metric d' coincides with d up to a constant factor $\alpha > 0$.

Proof:

(i) 1. $d(x,y) < \infty$ by (425)(iii).

2. $d(x,x) = 0$ because $\lambda(x,x,a) = 1$ and $\lambda(u,v,a) \to \infty$.

3. $d(x,y) = 0 \Rightarrow x = y$. Otherwise, in case of $x \neq y$, we would conclude $\lim\limits_{a \in R_o} \frac{\lambda(u,v,a)}{\lambda(x,y,a)} = \infty$

4. $d(x,y) = d(y,x)$ because a chain between x and y is a chain between y and x.

5. $d(x,z) \le d(x,y) + d(y,z)$. A chain between x and y of length λ_1 and a chain between y and z of length λ_2 (both of order a) can be combined to a chain between x and z of length $\lambda_1 + \lambda_2$. A minimal chain could be only shorter, hence $\lambda(x,z,a) \le \lambda(x,y,a) + \lambda(y,z,a)$. The assertion now follows by division by $\lambda(u,v,a)$ and taking limits.

(ii) Transport mappings map minimal chains onto minimal chains.

(iii) This follows from

$$\frac{\lambda(x,y,a)}{\lambda(u',v',a)} = \frac{\lambda(x,y,a)}{\lambda(u,v,a)} \frac{\lambda(u,v,a)}{\lambda(u',v',a)}$$

by taking limits. □

(4232) Definition:

(i) Let $M \subset P$, $x \in P$, $a \in \bar{R}_o$,

$$\lambda(x,M,a) \underset{\text{def}}{=} \inf\{\lambda(x,y,a) \mid y \in M\}.$$

The infimum is attained, since $\lambda(x,y,a)$ has only values in $\mathbb{N}$.

(ii) Let $M \subset P$ be compact, $x \in P$,

$$d(x,M) \underset{\text{def}}{=} \inf\{d(x,y) \mid y \in M\}.$$

The infimum is attained, since either $x \in M$ or $M \subset P \setminus f$, $f \in \tilde{R}_o$, $x \in f$, and $d(x,-)$ is continuous on $P \setminus f$ by (429).

(4233) Proposition: Let $x \in P$ and $M \subset P$ be compact.

(i) $d(x,M) = \lim\limits_{a\in\bar{R}_o} \frac{\lambda(x,M,a)}{\lambda(u,v,a)}$.

(ii) Let $x \notin M$, $M = \biguplus\limits_{j\in J} M_j$, where all M_j are compact. Then

$$\frac{\lambda(x,M_j,a)}{\lambda(u,v,a)} \xrightarrow[a\in\bar{R}_o]{} d(x,M_j)$$

uniformly w.r. to $j \in J$.

Before proving this proposition we consider the following

(4234) Lemma: Let I be a directed set and $(\lambda_i)_{i\in I}$ a family of functions $\lambda_i : X \to \mathbb{R}^+$, converging uniformly to a function $\lambda : X \to \mathbb{R}^+$. Let $X = \biguplus\limits_{j\in J} X_j$, J some index set, and assume that λ_i attains its infimum on X_j at the point $x_{ij} \in X_j$ and λ at the point $x_j \in X_j$. Then $\lim\limits_{i\in I} \lambda_i(x_{ij}) = \lambda(x_j)$ uniformly w.r. to $j \in J$.

Proof: Let $\delta > 0$ be given. $\exists\, i_o \in I\ \forall\, i > i_o\ \ \forall\, x \in X$ the following holds:

$|\lambda_i(x)-\lambda(x)| < \delta$, that is

$\lambda_i(x) - \lambda(x) < \delta$, $\lambda(x) - \lambda_i(x) < \delta$. In particular:

$\lambda_i(x_{ij}) - \lambda(x_j) \leq \lambda_i(x_j) - \lambda(x_j) < \delta$ and

$\lambda(x_j) - \lambda_i(x_{ij}) \leq \lambda(x_{ij}) - \lambda_i(x_{ij}) < \delta$. Hence:

$|\lambda_i(x_{ij})-\lambda(x_j)| < \delta$ □

Proof of (4233):

(i) We may assume $x \notin M$ and uniform convergence of the system of functions $\left(y \mapsto \frac{\lambda(x,y,a)}{\lambda(u,v,a)}\right)_{a\in\bar{R}_o}$ on M. Let the infimum $\lambda(x,M,a)$ be at $y_a \in M$ and the infimum $d(x,M)$ at y_m. By (4234),

$$\lim_{a\in\bar{R}_o} \frac{\lambda(x,y,a)}{\lambda(u,v,a)} = d(x,y_m) = d(x,M).$$

(ii) follows by analogous reasoning. □

Now, as in section 4.1, we consider arbitrarily small, compact, connected, invariant subgroups $N \subset \bar{T}$, such that $\bar{T}/N$ will be a Lie group. Let us recall the abbreviations

$\hat{T} = \bar{T}/N$, $\hat{J} = JN/N$, $\hat{P} = P/N$,

further we set

$\hat{R}_o \underset{def}{=} \{\hat{a} \mid \hat{a}=\{Nx \mid x\in a\}\subset\hat{P}, a\in\bar{R}_o\}$.

(4235) Theorem: The chain quotient w.r. to $(\hat{R},\subset,\hat{T})$ converges uniformly on regions (in the sense of (426)).

Proof: Let $\hat{a}$, $\hat{f} \in \hat{R}_o$ and $\hat{x} \in \hat{f}$ such that $x \in f$. We will show uniform convergence of the chain quotient on the subset $\hat{\bar{a}}\setminus\hat{f}$. Since $\hat{\bar{a}}\setminus\hat{f} = \{Ny \mid y\in A\}$, where $A \underset{def}{=} \bar{a}\setminus Nf$, we may identify $M = NA = \biguplus_{y\in A} Ny = \biguplus_{j\in J} M_j$ and apply (4233)(ii).

The uniform convergence now follows, if we can show that $\hat{\lambda}(\hat{x},\hat{y},\hat{a}) = \lambda(x,Ny,a)$. Let $[\hat{x},\hat{y},\hat{a},\hat{\tau}_1,..\hat{\tau}_\lambda]$ be a minimal chain w.r. to $(\hat{R},\subset,\hat{T})$. This means in particular:

$\hat{x} \in \hat{\tau}_1\hat{a} = \{\hat{\tau}_1\ Nx \mid x\in a\} = \{N\tau_1\ x \mid x\in a\}$, thus $\exists\ n_1 \in N$ such that $x \in n_1\ \tau_1[a]$. Further, $\hat{x}_1 \in \hat{\tau}_1\hat{a} \cap \hat{\tau}_2\hat{a} \neq \emptyset$ means that $\exists\ m_1, m_2 \in N$ such that $x_1 \in m_1\ \tau_1[a] \cap m_2\ \tau_2[a]$, hence

$n_1\ m_1^{-1}\ x_1 \in n_1\ \tau_1[a] \cap n_1\ m_1^{-1}\ m_2\ \tau_2[a]$. Now put

$n_2\ \tau_2 \underset{def}{=} n_1\ m_1^{-1}\ m_2\ \tau_2$ and continue until a chain $[x,x_\lambda,a,n_1\tau_1,...n_\lambda\tau_\lambda]$, where $x_n \in Ny$, is obtained. It is minimal because a shorter chain between x and some $\tilde{x} \in Ny$ would yield a shorter $\hat{a}$-chain between $\hat{x}$ and $\hat{y}$ if one considers the quotient w.r. to N, and this is contradictious to $\lambda = \hat{\lambda}(\hat{x},\hat{y},\hat{a})$. This shows $\hat{\lambda}(\hat{x},\hat{y},\hat{a}) = \lambda(x,Ny,a)$. □

(4236) Theorem: Let $\hat{d} : \hat{P} \times \hat{P} \to \mathbb{R}^+$ denote the chain distance according to (4235) ($\hat{u},\hat{v}\in\hat{P}$ are kept fixed).

(i) $\hat{d}(\hat{x},-) : \hat{a}\setminus\{\hat{x}\} \to \mathbb{R}^+$ is continuous for all $\hat{a} \in \hat{R}_o$.

(ii) Let us define $K_\leq(\hat{x},\varepsilon) = \{\hat{y}\in\hat{P} \mid \hat{d}(\hat{x},\hat{y})\leq\varepsilon\}$ (analogously: $K_<, K_=$). Then each neighbourhood U of $\hat{x}$ contains some ball $K_\leq(\hat{x},\varepsilon)$, $\varepsilon > 0$.

(iii) The uniformity induced by $\hat{d}$ coincides with the given uniformity on $\hat{P}$.

Proof:

(i) follows analogously to (429).

(ii) We may assume that U has a compact boundary ∂U which dissects $\hat{P}$. Let $V \supset U$ be a compact set. The chain quotient converges uniformly on $\overline{V \setminus U}$. Now assume the converse of (ii), namely that for each $\varepsilon > 0$ there exists some $\hat{y}_\varepsilon \notin U$ such that $\hat{d}(\hat{x},\hat{y}_\varepsilon) \le \varepsilon$. Let $a \in \hat{R}_o$ be a sufficiently small connected region (notice that $\hat{P}$ is locally connected) and consider a minimal a-chain between $\hat{x}$ and $\hat{y}_\varepsilon$. Its intersection with ∂U contains a points $\hat{z}_\varepsilon$ such that

$\left|\frac{\lambda(\hat{x},\hat{z}_\varepsilon,a)}{\lambda(\hat{u},\hat{v},a)} - d(\hat{x},\hat{z}_\varepsilon)\right| \le \varepsilon$. Since $\lambda(\hat{x},\hat{z}_\varepsilon,a) \le \lambda(\hat{x},\hat{y}_\varepsilon,a)$ and $d(\hat{x},\hat{y}_\varepsilon) \le \varepsilon$ we have $d(\hat{x},\hat{z}_\varepsilon) \le 3\varepsilon$. By compactness of ∂U we select a sequence $\varepsilon_i \to 0$ such that $\hat{z}_{\varepsilon_i} \to \hat{z} \in \partial U$, and, by continuity of $\hat{d}(\hat{x},-)$, $d(\hat{x},\hat{z}) = 0$. Since $\hat{x} \neq \hat{z}$ this is a contradiction.

(iii) By (i) and (ii) the two topologies coincide. In each case the corresponding uniformity is generated from the topology by the group $\hat{T}$. □

(4237) Corollary: Let $0 < \varepsilon$, $r_1 < r_2$ and $\hat{y} \in \hat{P}$ such that $r_1 \le d(\hat{x},\hat{y}) \le r_2$. Then $\exists\, b \in \hat{R}_o\ \forall\, a \in \hat{R}_o$ such that $a \subset b$ each point $\hat{z}$ on a minimal a-chain between $\hat{x}$ and $\hat{y}$ satisfies: $\hat{d}(\hat{x},\hat{z}) \le r_2 + \varepsilon$.

Proof: Choose $b \in \hat{R}_o$ such that $\sup\{\hat{d}(\hat{r},\hat{s}) \mid \hat{r},\hat{s} \in b\} < \varepsilon/3$ and the distance $\hat{d}(\hat{x},\hat{w})$ can be uniformly approximated up to $\varepsilon/3$ by each a-chain quotient, $a \subset b$, if $r_1 \le \hat{d}(\hat{x},\hat{w}) \le r_2 + \varepsilon$ (according to (4235)). Assume $d(\hat{x},\hat{z}) > r_2 + \varepsilon$.

Since a has a diameter less than $\varepsilon/3$, there exists a point $\hat{w}$ on the minimal a-chain between $\hat{x}$ and $\hat{y}$ such that $r_2 + \frac{2}{3}\varepsilon < \hat{d}(\hat{x},\hat{w}) \le r_2 + \varepsilon$.
Now

$$\hat{d}(\hat{x},\hat{w}) \le \frac{\hat{\lambda}(\hat{x},\hat{w},a)}{\hat{\lambda}(\hat{u},\hat{v},a)} + \varepsilon/3$$

$$\le \frac{\hat{\lambda}(\hat{x},\hat{y},a)}{\hat{\lambda}(\hat{u},\hat{v},a)} + \varepsilon/3$$

$$\le d(\hat{x},\hat{y}) + \frac{2}{3}\varepsilon$$

$$\le r_2 + \frac{2}{3}\varepsilon ,$$

which is a contradiction. □

Now consider the canonical chart around $\hat{x} \in \hat{P}$ (see section 4.1)

$f^{-1} : U \to K$

and the dilatations

$D_t : K \to K$

$Z \mapsto t^{-1}Z,\ t \in]0,1]$.

Each congruent mapping $\tau \in \hat{T}$ induces a local congruent mapping $D_t\, d^{-1}\, \tau\, f\, D_t^{-1}$, which operates in the 0-neighbourhood $f^{-1}[U]$ of K. Intuitively speaking, one looks at the congruent mappings in the canonical chart with a magnifying glass. One then expects the space to look almost Euclidean.
Indeed, by (4111) $f^{-1}\, \hat{J}\, f$ consists of local, orthogonal transformations commuting with dilations. It remains to examine transformations occuring in exp K. We set for $t \in]0,1]$ and $X \in K$

(4238) $\tau(t,X) \underset{def}{=} D_t\, f^{-1} \exp(tX)\, f\, D_t^{-1}$
and notice that the local transformations $\tau(t,X)$ will generally be non-linear. However, for $t \to 0$, they may be approximated by translations of K, as we will show.

(4239) Lemma: Let $X,Y,K \in K$ and $\alpha,\beta \in \mathbb{R}$ be such that the following expression are defined, and put $\tau(t,X,Y) \underset{def}{=} \tau(t,X)Y$.

(i) $K = \tau(t,X,Y) \Leftrightarrow \exp(tK)\hat{x} = \exp(tX)\ \exp(tY)\hat{x}$,

(ii) $\tau(t,X,0) = X$,

(iii) $\tau(t,\alpha X,\beta X) = (\alpha+\beta)X$,

(iv) $\tau(t,X)^{-1} = \tau(t,-X)$.

Proof:

(i) By (4110), $f(X) = (\exp X)\hat{x}$ and, by definition of $\tau(t,X)$,

$K = \tau(t,X,Y) \Rightarrow f(tK) = \exp(tX)\ f(tY) = \exp(tX)\ \exp(tY)\hat{x}$,

hence $\exp(tK)\hat{x} = \exp(tX)\ \exp(tY)\hat{x}$.

(ii),(iii) and (iv) are consequences of the following fact: Since f is locally diffeomorphic, $\exp(K_1)\hat{x} = \exp(K_2)\hat{x}$ implies $K_1 = K_2$, provided K_1 and K_2 are chosen from $f^{-1}[U] \subset K$. □

(4239)(i) yields a method of computing $K = \tau(t,X,Y)$. Find a $Z \in T$ satisfying

(42310) $\exp Z = \exp(tX)\ \exp(tY)$.

This can be accomplished by the Baker-Campbell-Hausdorff formula (see [VAR] 2.15.). $Z(X,Y)$ is analytic w.r. to (X,Y) and can be written as an absolutely convergent series

(42311) $Z = \sum_{h\geq 0} t^n c_n(X:Y)$,

where the c_n are polynomial mappings $T \times T \to T$ of degree n, for instance:

$c_o = 0$

$c_1 = X + Y$

$c_2 = \frac{1}{2}[X,Y]$

$c_3 = \frac{1}{12}[X,[X,Y]] - \frac{1}{12}[Y,[X,Y]]$

$\vdots$

Further we obtain a locally unique decomposition

(42312) $\exp Z = \exp K' \exp J'$, $K' \in K$, $J' \in J$ where the assignements $Z \mapsto K'$, $Z \mapsto J'$ are locally analytic, since the canonical chart and $\exp: T \to \hat{T}$ are analytic. If we set

(42313) $K' = tK$, $J' = tJ$

we have $K = \tau(t,X,Y)$ and

(42314) $K = \sum_{n\geq 0} t^n K_n(X,Y)$, $J = \sum_{n\geq 0} t^n J_n(X,Y)$, from which K_n can be computed for any power of t in the following manner.

Inserting into (42310) and using (42311) we obtain

$\exp(tX) \exp(tY) = \exp(tK) \exp(tJ)$ and

$$\sum_{n\geq 0} t^n c_n(X:Y) = \sum_{n\geq 0} t^n c_n(K:J)$$

$$= \sum_{n\geq 0} t^n c_n\Big(\sum_{\nu\geq 0} t^{\nu} K_{\nu}(X,Y) : \sum_{\mu\geq 0} t^{\mu} J_{\mu}(X,Y)\Big).$$

The power series are identical iff the t^m-coefficients coincide. Since $c_o = 0$, no t^o-terms occur. For the t^1-terms we obtain

$t^1\, c_1(X:Y) = t^1\, c_1(K:J)$ modulo t^m-terms, $m > 1$.

Thus only the t^o-portion of $c_1(K:J)$ counts, which is $K_o + J_o$.

Hence:

$t(X+Y) = t(K_o+J_o)$,

and, since $X + Y \in K$,

(42315) $K_o = X + Y$, $J_o = 0$.

The next terms are

$$K_1 = \tfrac{1}{2}[X,Y]_K,\quad J_1 = \tfrac{1}{2}[X,Y]_J$$

$$K_2 = \tfrac{1}{12}[X-Y,[X,Y]]_K - \tfrac{1}{2}[X+Y,[X,Y]_J]$$

$$J_2 = \tfrac{1}{12}[X-Y,[X,Y]]_J,$$

where the subscript $_K$ (resp. $_J$) denotes the projection onto the corresponding subspace. Further, $[K,J] \subset K$ is used. The calculations could be extended to obtain a B.C.H. formula for (reductive) homogeneous spaces.

Now we consider the norm of

$$\frac{K-K_o}{t} = \sum_{n\geq 1} t^{n-1} K_n \underset{def}{=} L(t,X,Y), \quad t \in]0,1].$$

Since L is analytic in (t,X,Y), it has a bounded derivative at (0,0,0), hence

$$\|K(t,X,Y)-K_o(X,Y)\| = t\|L(t,X,Y)\| \leq t(\|L(0)\|+|t|\,\|X\|\,\|Y\|\,\|L_1\|),$$

where

$$\|L(0)\| = \|K_1\| = \|\tfrac{1}{2}[X,Y]_K\| \leq C\|X\|\,\|Y\|$$

and t,X,Y are sufficiently small.

We recall $K(t,X,Y) = \tau(t,X,Y)$ and $K_o(X,Y) = X + Y$ and obtain:

(42316) Lemma: There exists a connected 0-neighbourhood $\mathcal{W} \subset f^{-1}[U] \subset K$ such that

$\forall \tilde{\varepsilon} > 0 \ \exists t_o > 0 \ \forall t \in]0,t_o[\quad \forall X, Y \in \mathcal{W}$,

$\|\tau(t,X,Y)-(X+Y)\| < \tilde{\varepsilon}\|X\|\,\|Y\|$.

Let us summarize: The group $\hat{T}$ operates on K by means of the class of local mappings $T^{(t)} \underset{def}{=} D_t f^{-1} \hat{T} f D_t^{-1}$. Each local transformation $\tau^{(t)} \underset{def}{=} D_t f^{-1} \tau f D_t^{-1} \in T^{(t)}$ can be written as a product $\tau^{(t)} = \tau(t,X)\, j$ such that $X = \tau^{(t)}0$ and $j \in f^{-1} \hat{J} f \underset{def}{=} J'$. Each $\tau \in T^{(t)}$ has a domain of definition containing $\mathcal{D}(\tau) \underset{def}{=} \tau^{-1}[\mathcal{W}] \cap \mathcal{W}$.

If we consider the class of regions $R' \underset{def}{=} \{a' | a'=f^{-1}[a]$, where $a\in\hat{R}, a\subset U$ and $0\in a'\}$ in K, it is appropriate to speak of $(R',\subset,T^{(t)})$-chains between points in $\mathcal{W}$. What can be said about the convergence of the chain quotient? Notice that a minimal $(R',\subset,T^{(t)})$- chain between 0 and $Y \in \mathcal{W}$, say $[0,Y,a',\tau_1^{(t)},\ldots\tau_\lambda^{(t)}]$, is just the image of a minimal $(\hat{R},\subset,\hat{T})$-chain

$[\hat{x}=f\circ D_t^{-1}(0), f\circ D_t^{-1}(Y), f\circ D_t^{-1}[a'], \tau_1,\ldots\tau_\lambda]$,

provided that $a' \subset \mathcal{D}(\tau_i^{(t)}))$ for $i=1\ldots\lambda$.

But this can be achieved by (4237), if a' is sufficiently small (not necessarily independent of t). Let us make the following

(42317) Definition: $S_{\leq}(X,\varepsilon) \underset{\text{def}}{=} \{Y\in K \mid \|X-Y\|\leq\varepsilon\}$, analogously: $S_<$ and $S_=$.

Then we can state the

(42318) Proposition: $\exists\ r > 0$ such that $S_{\leq}(O,r) \subset W$ and

$\forall\ V,Y \in S_=(O,r)\ \forall\ t \in\ ]0,1[$

$$\lim_{O\in R'} \frac{\lambda^{(t)}(O,Y,a)}{\lambda^{(t)}(O,V,a)}$$ exists and will be denoted by $d^{(t)}(O,Y)$.

Proof: Let ε, r_2 in (4237) be such that $K_{\leq}(\hat{x},r_2+\varepsilon) \subset W$ and choose $r > 0$ satisfying $f[S_{\leq}(O,r)] \subset K_{\leq}(\hat{x},r_2)$.

This is possible by (4236)(iii). Now apply (4237) for

$r_1 \underset{\text{def}}{=} \hat{d}(\hat{x}, f\circ D_t^{-1}Y) < r_2$.

Thus each point $\hat{w}$ on a minimal $\hat{a}$-chain, provided $\hat{a} \subset \hat{b}$, between $\hat{x}$ and $\hat{y} = f \circ D_t^{-1}\ Y$ lies in $K_{\leq}(\hat{x},r_2+\varepsilon)$, hence in W. Therefore $D_t \circ f^{-1}$ maps this minimal chain onto a minimal $(R',\subset,T^{(t)})$-chain between O and Y. This proves the assertion. □

(42319) Proposition: Let $V,Y \in K$ satisfy $\|V\| = \|Y\| = r$.

$\forall\ \varepsilon \in\ ]0,1[\ \exists\ t_o \in\ ]0,1[\ \forall\ t \in\ ]0,t_o[$ the following 2 assertions hold:

(i) $\forall\ a \in R'\ \forall\ \tau \in T^{(t)},\ \|\tau(O)\| \leq 4r \Rightarrow \Delta(\tau[a]) \leq (1+\varepsilon)\ \Delta(a)$.

(ii) If $a_n \underset{\text{def}}{=} S_<\left(O, \frac{r}{2n}\ (1+\frac{1}{n})\right)$, then $\left|\lim_{n\to\infty} \frac{\lambda^{(t)}(O,Y,a_n)}{\lambda^{(t)}(O,V,a_n)} - 1\right| < \varepsilon$.

Proof:

(i) Apply (42316) for $\tilde{\varepsilon} \leq \varepsilon/8r$ in order to find $t_o \in\ ]0,1[$. Take any $X,Z \in a$ and write $\tau = \sigma \circ j$, where $\sigma = \tau(t,U)$ and $U = \tau(O)$.

$$\|\tau X-\tau Z\| \leq \|\sigma jX-(U+jX)\| + \|\sigma jZ-(U+jZ)\| + \|(U+jX)-(U+jZ)\|$$

(by (42316)) $\leq \tilde{\varepsilon}\|U\|\,\|jX\| + \tilde{\varepsilon}\|U\|\,\|jZ\| + \|X-Z\|$

(since $O\in a$) $\leq \Delta(a)\ (2\tilde{\varepsilon}\|U\| + 1)$

$\leq \Delta(a)\ (2\cdot\tilde{\varepsilon}\cdot 4r + 1)$

$\leq (1+\varepsilon)\ \Delta(a)$.

(ii) Let $\varepsilon \in]0,1[$ be given and chose $t_o \in]0,1[$ as in part (i). It is easily shown that

$[0,V,a_n,\tau(t,\frac{1}{2n} V),\tau(t,\frac{2}{2n} V),\ldots (t,\frac{2n-1}{2n} V)]$ is a chain because $\frac{i}{n} V = \tau(t,\frac{2i-1}{2n} V) \frac{1}{2n} V = \tau(t,\frac{2i+1}{2n} V)(- \frac{1}{2n} V)$ by (4237)(iii).

Let $[\hat{x},f\circ D_t^{-1}(V),f\circ D_t^{-1}(a_n),\tau_1,\ldots\tau_\lambda]$ be a minimal $(\hat{R},\subset,\hat{T})$-chain.

Clearly $\lambda \leq n$.

Let $[0,V,a_n,\tau_1^{(t)},\ldots\tau_\lambda^{(t)}]$ be the corresponding $(R',\subset,T^{(t)})$-chain.

We infer:

$$1 \geq \frac{\lambda^{(t)}(0,V,a_n)}{n} \geq \frac{r}{n \sup\{\Delta(\tau^{(t)}[a_n])\,|\,j=1\ldots\lambda\}} \geq \frac{r}{n(1+\varepsilon)\frac{r}{n}(1+\frac{1}{n})} = \frac{n+1}{n(1+\varepsilon)} .$$

The same argument applied to Y yields:

$$1 \geq \frac{\lambda^{(t)}(0,Y,a_n)}{n} \geq \frac{n+1}{n(1+\varepsilon)} .$$

Combining the inequalities we obtain

$$\frac{n+1}{(1+\varepsilon)n} \leq \frac{\lambda^{(t)}(0,Y,a_n)}{\lambda^{(t)}(0,V,a_n)} \leq \frac{n(1+\varepsilon)}{n+1} .$$

Now the assertion follows if $n \to \infty$. □

We are now prepared to show that for some $V \in S_=(0,r)$ the orbit $J'V$ dissects the 0-neighbourhood $\mathcal{W}$. The idea of the proof has been portrayed in the introduction (see fig. (131)).

(42320) Theorem: There exists a $V \in K$ such that $\|V\| = r$ and $\mathcal{W}\setminus J'V$ is not connected. (We recall that $\mathcal{W}$ was assumed to be connected.)

Proof:

1. Let $V \in K$ and $\|V\| = r$. $J'V$ is a closed subset of $S_=(0,r)$, the latter dissecting $\mathcal{W}$. In order to derive a contradiction let us assume

(1) $J'V \subsetneqq S_{=}(O,r)$.

Let $<,>$ denote the inner product on K. Since $J'V$ is closed and $\forall X,Y \in S_{=}(O,r)$, $<X,Y> = r^2 \Leftrightarrow X = Y$, we may conclude:

(2) $\exists Y \in S_{=}(O,r) \setminus J'V \;\; \exists \delta \in]O,1[\;\; \forall X \in J'V, <X,Y> \leq \delta r^2$.

2. Let $n \in \mathbb{N}$, $t \in]O,1[$, $\varepsilon_n > O$ be for the moment arbitrarily chosen. We consider chains $[O,V,a_n,\tau_o \ldots \tau_{n-1}]$, where $a_n = b_n \cup c_n$,

(3) $b_n \subset S_<(O,\varepsilon_n)$,

(4) $c_n \subset S_<(\frac{1}{n} V,\varepsilon_n)$,

and $\tau_i = \tau(t,\frac{i}{n}V)$ for $i=0 \ldots n-1$.

The chain property is easily proved analogously to (42319)(ii). A minimal chain could only be shorter, hence

(5) $\lambda^{(t)}(O,V,a_n) \leq n$.

3. Now consider $\delta > O$ according to (2) and choose $\varepsilon > O$ such that

(6) $\varepsilon < \frac{1-\delta}{2} < \frac{1}{2}$.

We set $\tilde{\varepsilon} = \varepsilon/8r$ and choose $t_o > O$ according to (42316). Thus, we may also use (42319). Let $t \leq t_o$ be fixed. For each integer $n \in \mathbb{N}$ we choose some $\varepsilon_n > O$ satisfying

(7) $\varepsilon_n < \frac{3r}{2(1+\varepsilon)} - \frac{r}{2n}$ $\quad$ ($> O$ since $3n > 1 + \varepsilon$), hence:

(8) $\varepsilon_n < \frac{4r}{\varepsilon} - \frac{r}{4n} < \frac{8r}{\varepsilon}$.

Further, ε_n is chosen so that

(9) $\varepsilon_n < \frac{1}{4n^2(1+\varepsilon)}$,

(10) $\varepsilon_n < \frac{r}{2(1+\varepsilon)^2}$ $\quad$ and

(11) $\varepsilon_n < \frac{r}{2n}\left[\frac{3}{(1+2\varepsilon)(1+\varepsilon)} - 1\right]$ $\quad$ ($> O$ since $\varepsilon < \frac{1}{2}$).

Consider $a_n = b_n \cup c_n$ according to (3) and (4). By (42318) and (42319)(ii),

$$\lim_{n\to\infty} \frac{\lambda^{(t)}(O,Y,a_n)}{\lambda^t(O,V,a_n)} = d^{(t)}(O,Y) < 1 + \varepsilon, \text{ hence } \exists\, n_o \in \mathbb{N}\ \forall\, n \geq n_o$$

$\lambda^{(t)}(O,Y,a_n) < \lambda^{(t)}(O,V,a_n)(1+2\varepsilon)$ and by (5):

(12) $\lambda^{(t)}(O,Y,a_n) < n(1+2\varepsilon)$.

From (2) it follows that,

(13) $\forall\, X \in J'(\frac{1}{n} V)$, $\langle Y,X\rangle \leq \delta r^2/n$.

4. Now consider a minimal chain $[O,Y,a_n,\tau_1 \dots \tau_\lambda]$.

We will always assume $n \geq n_o$, further

(14) $n^2 \geq \frac{2+\varepsilon}{16r}$.

This implies:

$$\frac{8r}{2+\varepsilon} \geq \frac{1}{2n^2} \underset{(9)}{>} \varepsilon_n$$

$4r \geq \varepsilon_n + \frac{1}{2}\varepsilon\, \varepsilon_n$

$4r(1-\varepsilon\, \varepsilon_n/8r) \geq \varepsilon_n$, hence:

(15) $4r \geq \dfrac{\varepsilon_n}{1-\varepsilon\, \varepsilon_n/8r}$,

where the denominator is strictly positive by virtue of (8).

4.1. We will prove the following auxiliary assertion by induction on i:

$\forall\, i=1\dots\lambda\ \forall\, W \in \tau_i[a_n]$,

(16) $\langle Y,W\rangle < ir[r(\delta+\varepsilon)/n+4\varepsilon_n(1+\varepsilon)]$ and

(17) $\|W\| \leq 3r$.

<u>4.2. "i=1".</u>

If $U_1 \underset{def}{=} \tau_1(O)$ we have

(18) $\tau_1 = \tau(t,U_1) \circ j_1$, $j_1 \in J'$.

We may well confine ourselves to the case $O \in \tau_1[b_n]$ and $W \in \tau_1[c_n]$. Let $W = \tau_1 W_o$, $W_o \in c_n$, and $W_1 \underset{def}{=} j_1 W_o$. Hence

$W = \tau(t,U_1)W_1$ and

(19) $\|W_1\| = \|W_o\| \le \frac{r}{n} + \varepsilon_n$.

From (13) we conclude

(20) $\langle Y, j_1(\frac{1}{n} V)\rangle \le \delta r^2/n$, further $\|\frac{1}{n} V - W_o\| < \varepsilon_n$ since $W_o \in c_n$ and, because j_1 is a $\|.\|$-isometry,

(21) $\|j_1(\frac{1}{n} V) - W_1\| < \varepsilon_n$.

4.2.1. We define $W_2 \underset{def}{=} U_1 + W_1$ and will show:

(22) $\|W_1 - W_2\| = \|U_1\| < \dfrac{\varepsilon_n}{1-\varepsilon_n\ \varepsilon/8r}$.

Proof: We conclude:

$0 \in \tau_1[b_n]$

$Z_1 \underset{def}{=} \tau_1^{-1}(0) \in b_n$ and $\|Z_1\| < \varepsilon_n \quad \|j_1(Z_1)\| < \varepsilon_n$

$$\begin{aligned}\|U_1\| = \|U_1 - \tau_1 Z_1\| &\le \|U_1 - (U_1 + j_1 Z_1)\| + \|(U_1 + j_1 Z_1) - \tau(t,U_1) j_1 Z_1\| \\ &\underset{(42316)}{\le} \|j_1 Z_1\| + \|U_1\| \cdot \|j_1 Z_1\| \cdot \varepsilon/8r \\ &\le \varepsilon_n + \|U_1\|\ \varepsilon_n\ \varepsilon/8r.\end{aligned}$$

From this (22) follows. □

4.2.2. Now we infer

$$\|W - W_2\| = \|\tau(t,U_1)W_1 - (U_1 + W_1)\| \underset{(42316)}{\le} \|U_1\|\ \|W_1\|\ \varepsilon/8r$$

and by virtue of (22) and (19):

(23) $\|W_2 - W\| \le \dfrac{\varepsilon_n}{1-\varepsilon_n\ \varepsilon/8r}\ (r/n+\varepsilon_n)\ \dfrac{\varepsilon}{8r}$.

We combine (21), (22) and (23) concluding:

$$\begin{aligned}\|j_1(\tfrac{1}{n} V) - W\| &\le \|j_1(\tfrac{1}{n} V) - W_1\| + \|W_1 - W_2\| + \|W_2 - W\| \\ &\le \varepsilon_n + \frac{\varepsilon_n}{1-\varepsilon_n\ \varepsilon/8r}\left[1 + \frac{r\varepsilon + n\ \varepsilon_n\ \varepsilon}{8nr}\right] \\ &= \varepsilon_n\left[1 + \frac{8r}{8r-\varepsilon_n\ \varepsilon} \cdot \frac{8nr + r\varepsilon + n\ \varepsilon_n\ \varepsilon}{8nr}\right] = \varepsilon_n\ \frac{r(16n+\varepsilon)}{n(8r-\varepsilon_n\ \varepsilon)}\end{aligned}$$

$$\underset{(8)}{\leq} \varepsilon_n \frac{r(16n+\varepsilon)}{n(8r-r(4-\frac{\varepsilon}{4n}))} = 4\ \varepsilon_n$$

Together with (20), this shows

$$\langle Y,W\rangle \leq \langle Y,j_1(\tfrac{1}{n}\ V)\rangle + \|Y\|\,\|W-j_1(\tfrac{1}{n}\ V)\| \leq \delta r^2/n + 4r\ \varepsilon_n$$

$$\leq r[r(\delta+\varepsilon)/n+4\ \varepsilon_n(1+\varepsilon)]$$

which is (16) for $i = 1$.

4.2.3. Combining (22) and (15) we infer

(24) $\|\tau_1(O)\| = \|U_1\| < 4r$

and may apply (42319)(i). Hence

$$\Delta(\tau_1[a_n]) \leq (1+\varepsilon)\ \Delta(a_n) \leq (1+\varepsilon)\ (\tfrac{r}{n}+2\ \varepsilon_n)$$

and thus

$$\|W-O\| \leq (1+\varepsilon)\ (\tfrac{r}{n}+2\ \varepsilon_n) \underset{(9)}{\leq} (1+\varepsilon)(\tfrac{r}{n}+\tfrac{3r}{1+\varepsilon}-\tfrac{r}{n}) = 3r.$$

This proves (17) for $i = 1$.

<u>4.3. "i ⇒ i+1".</u>

Assume:

$W \in \tau_{i+1}[c_n], U_{i+1} = \tau_{i+1}(O),\ W_i \in \tau_i[c_n] \cap \tau_{i+1}[b_n] \neq \emptyset,$

$\tau_{i+1} = \tau(t,U_{i+1})j_{i+1}$ where $j_{i+1} \in J'$,

$W = \tau_{i+1} W_o,\ W_o \in c_n,$

$W_1 = j_{i+1} W_o,\ W_2 = W - U_{i+1}.$

The following holds:

(25) $\|W_1\| = \|W_o\| < \frac{r}{n} + \varepsilon_n,$

$$\langle Y,W_1\rangle = \langle Y,j_{i+1} W_o\rangle \leq \langle Y,j_{i+1}(\tfrac{1}{n}\ V)\rangle + \|Y\|\ \|W_o-\tfrac{1}{n}\ V\|$$

$$\underset{(13)}{\leq} \delta r^2/n + r\ \varepsilon_n, \text{ hence}$$

(26) $\langle Y,W_1\rangle \leq r(r\delta/n+\varepsilon_n).$

4.3.1. Let $W_i = \tau_{i+1} V_i$, $V_i \in b_n$, and $\tau_{i+1} = \sigma \circ \hat{\tau}$, where $\sigma = \tau(t,W_i)$ and $\hat{\tau} = \sigma^{-1} \circ \tau_{i+1}$. From $\hat{\tau} V_i = 0$ we infer $0 \in \hat{\tau}[b_n]$ and thus we may apply the results of 4.2. concerning τ_1 to $\hat{\tau}$. In particular, $\|\hat{\tau}(0)\| < 4r$ by virtue of (24), and $\|\sigma(0)\| = \|W_i\| \leq 3r$ by the induction hypothesis (17). Thus we may apply (42319)(i) for both $\hat{\tau}$ and σ and obtain

$$\Delta(\tau_{i+1}[b_n]) = \Delta(\sigma\circ\hat{\tau}[b_n]) \leq (1+\varepsilon)\ \Delta(\hat{\tau}[b_n]) \leq (1+\varepsilon)^2\ \Delta(b_n)$$

$$\leq (1+\varepsilon)^2\ 2\ \varepsilon_n. \quad \text{Hence}$$

$$\|U_{i+1}\| \leq \|W_i\| + \|U_{i+1}-W_i\| \leq 3r + 2(1+\varepsilon)^2\ \varepsilon_n \underset{(10)}{<} 4r,$$

(27) $\|U_{i+1}\| < 4r.$

Therefore, (42319)(i) applies to τ_{i+1}, as well as to τ_j, $j=1\ldots i$, and the triangle inequality yields:

$$\|W\| \leq (i+1)(1+\varepsilon)\ \Delta(a_n)$$

$$\leq \lambda(1+\varepsilon)(\tfrac{r}{n} + 2\ \varepsilon_n)$$

$$\underset{(12)}{<} n(1+2\varepsilon)(1+\varepsilon)(\tfrac{r}{n} + 2\ \varepsilon_n)$$

$$\underset{(11)}{<} 3r.$$

This proves (17).

4.3.2. From (16) as the induction hypothesis, it follows that,

(28) $\langle Y,W_i\rangle < ir[r(\delta+\varepsilon)/n+4\varepsilon_n(1+\varepsilon)].$

By (42319)(i) applied to τ_{i+1} we obtain

(29) $\|W_i-U_{i+1}\| = \|\tau_{i+1}V_i-\tau_{i+1}0\| \leq (1+\varepsilon)\ \varepsilon_n.$

Further:

$$\|W_2-W_1\| = \|W-U_{i+1}-W_1\| = \|\tau(t,U_{i+1})W_1-(U_{i+1}+W_1)\|$$

$$\underset{(42316)}{\leq} \|U_{i+1}\|\ \|W_1\|\ \varepsilon/8r$$

$$\underset{(25),(27)}{\leq} 4r \cdot (\tfrac{r}{n} + \varepsilon_n) \cdot \varepsilon/8r, \quad \text{hence,}$$

(30) $\|W_1-W_2\| \leq \frac{\varepsilon}{2}(\frac{r}{n}+\varepsilon_n)$.

We combine these results to compute an upper bound for $\langle Y,W\rangle$.

$$\begin{aligned}\langle Y,W\rangle &= \langle Y,U_{i+1}\rangle + \langle Y,W_2\rangle \\ &\leq \langle Y,W_i\rangle + \|Y\|\ \|W_i-U_{i+1}\| + \langle Y,W_1\rangle + \|Y\|\ \|W_1-W_2\| \\ &\leq ir[r(\delta+\varepsilon)/n+4\varepsilon_n(1+\varepsilon)] \quad \text{(by (28))} \\ &+ r(1+\varepsilon)\ \varepsilon_n \quad \text{(by (29))} \\ &+ r(r\delta/n+\varepsilon_n) \quad \text{(by (26))} \\ &+ r\ \frac{\varepsilon}{2}(\frac{r}{n}+\varepsilon_n) \quad \text{(by (30))} \\ &< (i+1)\ r[r(\delta+\varepsilon)/n+4\varepsilon_n(1+\varepsilon)].\end{aligned}$$

This completes the proof of (16). □

4.4. The chain property of $[0,Y,a_n,\tau_1,\ldots\tau_\lambda]$ implies $Y \in \tau_\lambda[a_n]$ and by (16):

(31) $\langle Y,Y\rangle = r^2 < \lambda r[r(\delta+\varepsilon)/n+4\varepsilon_n(1+\varepsilon)]$.

By (5), $\lambda_1 \underset{\text{def}}{=} \lambda^{(t)}(0,V,a_n) \leq n$, hence

$$\frac{\lambda}{\lambda_1} \geq \frac{\lambda}{n} \underset{(31)}{>} \frac{r}{r(\delta+\varepsilon)+4n\ \varepsilon_n(1+\varepsilon)}.$$

Taking the limit $n \to \infty$ the left hand side tends towards $d^{(t)}(0,Y)$ and the right towards $\frac{1}{\delta+\varepsilon}$ because $4n\ \varepsilon_n(1+\varepsilon) < \frac{1}{n}$ by (9). Hence,

(32) $d^{(t)}(0,Y) \geq \frac{1}{\delta+\varepsilon}$.

5. By (42319)(ii) we have $d^t(0,Y) < 1+\varepsilon$, thus (32) implies:

$$\begin{aligned}&\frac{1}{\delta+\varepsilon} < 1+\varepsilon \\ &1 < (1+\varepsilon)(\delta+\varepsilon) \\ &\underset{(6)}{<} (1+\frac{1-\delta}{2})(\delta+\frac{1-\delta}{2}) = \frac{3+2\delta-\delta^2}{4} \\ &4 < 3+2\delta-\delta^2 \\ &1 < \delta(2-\delta) = 1-(1-\delta)^2 < 1.\end{aligned}$$

This is a contradiction. □

The existence of dissecting orbits $\hat{J}\hat{y}$ in $\hat{P} = P/N_U$ for each $U \in \mathcal{U}$ completes the proof of "(iii) ⇒ (i)" because only this property was used in the section "(i) ⇒ (ii)" to show that $\bar{T}$ is a Lie group and thus $P = \hat{P}$.

4.3. TITS/FREUDENTHAL CLASSIFICATION

If the preceding axioms R1 to R6 hold, the resulting geometries $(P,\bar{T})$ can be exhaustively enumerated. We will adopt them from [FRE], but considering only connected groups $\bar{T}$.

The proof of this classification is much involved and it is no place here to give an account of it. Before enumerating the list of possible geometries we need some further definitions.

As usual, we denote by

$\mathbb{R}$ the field of real numbers,

$\mathbb{C}$ the field of complex numbers,

$\mathbb{Q}$ the (scew) field of quaternions,

$\mathbb{H}$ the alternative algebra of Cayley numbers (or octaves);

$\mathbb{F}$ $\mathbb{R}$ or $\mathbb{C}$ or $\mathbb{Q}$.

S^n is the n-dimensional real sphere and

P^n the n-dimensional projective real space, i.e. $P^n = S^n/\{\pm 1\}$.

We will use the notation of the classical groups, which is familiar to physicists, as for example $SO(n)$ or $U(n,\mathbb{F})$ (see for instance [GIL]). Further we will consider some groups corresponding to the exceptional simple complex Lie algebras G_2 and F_4. The different real forms may be characterized by the signature of the Cartan-Killing metric which will be indicated in brackets. We only need to consider the real groups $F_{4(-52)}$ and $F_{4(-20)}$ and $B_{4(-36)}$, which is the abstract group usually represented as $SO(9)$.

$\nu(G_2)$ denotes the 7-dimensional real representation of G_2 as the automorphism group of non-real Cayley numbers, which operates

transitively on S^6, hence on P^6. Further, let ν SO(2l+1) be the 2^l-dimensional real spin-representation of the special orthogonal group (more precisely: of its universal covering group Spin(2l+1)). In the case l = 3 its Lie algebra representation may be given by using a suitable basis of $\mathbb{H}$, $(e_i)_{i=0...7}$:

$(a_{ij}) \mapsto (x \in \mathbb{H} \mapsto \sum_{i,j=1}^{7} a_{ij} e_i (e_j x))$.

Hence ν SO(7) operates transitively on S^7 and P^7.
For all details see [FRE],[FRE3] and [F&V].

We are now prepared to formulate the list of possible geometries. They will be indicated either in the form

$P \leftarrow \bar{T}$ or

$P \leftarrow\!o\ J$ if P is an abelian subgroup and $\bar{T} = P \otimes J$ (semi-direct product),

or

$\frac{T_1}{J_1}$, where J_1 is a group which may be embedded into the group T_1 in a (more or less) obvious manner. In the third case, $P = \frac{T_1}{J_1}$ and $\bar{T}$ is equal to T_1 modulo the kernel of the natural operation of T_1 on P.
We will invent certain denotations for the non-classical geometries.

(431) Theorem: If R1 to R6 hold, $(P,\bar{T})$ is isomorphic (more precisely: $\sum_{TF}$-isomorphic, see section 5.3) to one of the following geometries:

(i) $\mathbb{F}^n \leftarrow\!o\ V(n)$, where V(n) is equal to
$\mathbb{F} = \mathbb{R}$: SO(n) or
$\mathbb{F} = \mathbb{C}$: U(n,$\mathbb{C}$) or SU(n,$\mathbb{C}$) or
$\mathbb{F} = \mathbb{Q}$: U(n,$\mathbb{Q}$) $\times$ K, where K is the group of right multiplications in $\mathbb{Q}^n$ with numbers of the form
$b \in \mathbb{Q}$, $|b| = 1$ or
$b \in \mathbb{C}$, $|b| = 1$ or
$b = 1$.

A geometry of this kind is called an "affine $\mathbb{F}$-Hilbert space

(or parabolic) geometry".

(ih) $\mathbb{H}^2 \leftarrow o\ \nu\, SO(9)$,

"parabolic octavian plane".

(is) $\mathbb{H} \leftarrow o\ \nu\, SO(7)$,

"octavian spin line".

(io) $\mathbb{R}^7 \leftarrow o\ \nu(G_2)$,

"imaginary octavian line".

(ii) $\dfrac{U(n,1;\mathbb{F})}{U(n,\mathbb{F}) \times U(1,\mathbb{F})}$, "$\mathbb{F}$-hyperbolic geometry".

(iih) $\dfrac{F_{4(-20)}}{B_{4(-36)}}$, "hyperbolic octavian plane".

(iii) $\dfrac{U(n+1;\mathbb{F})}{U(n,\mathbb{F}) \times U(1,\mathbb{F})}$, $(n \geq 2$ for $\mathbb{F} = \mathbb{R})$

"$\mathbb{F}$-elliptic geometry".

(iiih) $\dfrac{F_{4(-52)}}{B_{4(-36)}}$, "elliptic octavian plane".

(iiis) $P^7 \leftarrow \nu\, SO(7)$,

"elliptic spin line".

(iiio) $P^6 \leftarrow \nu(G_2)$,

"elliptic imaginary line".

(iv) $\dfrac{SO(n+1)}{SO(n)} = S^n$, $n \geq 2$, "spherical geometry".

(ivs) $S^7 \leftarrow \nu\, SO(7)$,

"spherical spin line".

(ivo) $S^6 \leftarrow \nu(G_2)$,

"spherical imaginary line".

5. CHARACTERIZATION OF EUCLIDEAN GEOMETRY

The classification of section 4.3 shows that Euclidean geometry could be singled out by two additional postulates:

(i) Vanishing curvature and

(ii) dimension 3.

It would be desirable to formulate these postulates in terms of $(R,<,T)$ in order to stress their empirical meaning. But this has only been achieved with respect to the axiom of dimension.

Finally we will study the class of Euclidean representations obtained in this way.

5.1 DIMENSION

It will be natural to employ a notion of dimension based on coverings.

(511) Definition: Let $b \in R_o$, $V \subset R$.

$$M(b,V) \underset{\text{def}}{\Leftrightarrow} \exists\, u \in R_o\ \forall\, v \in V\ \exists\, \tau \in T$$

such that $u < \tau b$ and $\tau b \wedge v \neq O$.

(V is b-approximately overlapping.)

(512) Lemma: If V is finite, $M(b,V) \Leftrightarrow \bigcap_{v \in V} N_b \tilde{v} \neq \emptyset$.

Proof: "⇒" Let $x \in \tilde{u}$, then $x \in N_b \tilde{v}$ for all $v \in V$.

"⇐" $\bigcap_{v \in V} N_b \tilde{v}$ is non-empty and open, hence contains a region $\tilde{u} \in \tilde{R}_o$. □

The following definition only presupposes the axioms R1 and R2.

(513) Definition: Let $N \in \mathbb{N} \cup \{-1,0\}$.

(i) $\underline{\text{Dim } R \leq N} \underset{\text{def}}{\Leftrightarrow}$

$\forall\, a \in R\ \forall\, v \in R_o\ \exists\, n \in \mathbb{N}\ \exists\, v_o \ldots v_n \in R$ satisfying:

$\forall\, i=1\ldots n\ \exists\, \sigma \in T$ such that $v_i < \sigma v$, $a < (v_1 \vee v_2 \vee \ldots v_n)$ and

$\forall\, V \subset \{v_1 \ldots v_n\}$, $M(v_o,V) \Rightarrow |V| \leq N + 1$.

(ii) $\underline{\text{Dim } R = N} \underset{\text{def}}{\Leftrightarrow}$ Dim $R \leq N$, but not Dim $R \leq N - 1$.

Thus it is required that each region $a \in R$ can be covered by a finite number of arbitrary small regions v_i such that at most $N + 1$ ones are "overlapping" (in the sense of (511)).

(514) Proposition: If $\mathrm{Dim}\, R \leq N$, then the following holds:

$$\forall \tilde{a} \in \tilde{R}\ \forall \tilde{v} \in \tilde{R}_o\ \exists n \in \mathbb{N}$$

$$\exists \tilde{v}_o \ldots \tilde{v}_n \in \tilde{R} \text{ satisfying}$$

$$\forall i = 0 \ldots n\ \exists \sigma \in T \text{ such that } \tilde{v}_i \subset \sigma$$

$$\tilde{a} \subset \bigcup_{i=1\ldots n} N_{v_o}\, v_i \text{ and}$$

$$\forall V \subset \{\tilde{v}_1, \ldots \tilde{v}_n\},\ \bigcap_{\tilde{v}_i \in V} N_{v_o}\, \tilde{v}_i \neq \emptyset \Rightarrow |V| \leq N + 1.$$

Proof: Since (514) is almost completely a $\sim$-version of (513)(i) we only need to prove an implication at the two points of deviation. One follows from (512), the other is of the form:

$$a < \bigvee_{i=1\ldots n} v_i \Rightarrow \tilde{a} \subset \bigcup_{i=1\ldots n} N_{v_o}\, \widetilde{v_i}.$$

By (336)(i),(v), $\tilde{a} \subset \widetilde{\bigvee_{i=1\ldots n} v_i} \sim \bigcup_{i=1\ldots n} \widetilde{v_i}$. But (335) implies

$$\widetilde{\bigvee_{i=1\ldots n} v_i} \subset N_{v_o}\left(\bigcup_{i=1\ldots n} \widetilde{v_i}\right) = \bigcup_{i=1\ldots n} N_{v_o}\, \tilde{v}_i \quad \text{for any } v_o \in R_o. \qquad \square$$

Consider $v, a, b \in R_o$, $\varepsilon > 0$ such that $\tilde{b} \subset \bar{b} \subset \tilde{a}$ and the diameter of $\tilde{v}$ being smaller than $\varepsilon/2$. (Recall that according to 4.2 P is endowed with a metric.) Further we will assume $\mathrm{Dim}\, R \leq N$. According to (514) the compact set $\bar{\tilde{b}}$ has finite, open coverings with mesh $\leq \varepsilon$, such that at most $N + 1$ members of the covering intersect. Then [H&W], Theorem V8 Cor., shows that $\bar{\tilde{b}}$ has dimension $\leq N$ (in the sense defined there). Since $\bar{\tilde{b}}$ contains a nonempty, open subset of P, namely $\tilde{b}$, P has dimension $\leq N$ as a manifold (see [H&W], Theorem IV 3, Cor. 1). These arguments show that we will single out just the 3-dimensional manifolds P in Freudenthal's list, if we postulate:

(515) <u>Axiom R7</u>: $\mathrm{Dim}\, R = 3.$

5.2 CURVATURE

We will prefer expressing the flatness of P not by the parallel axiom of Euclid but by a local property of triangles, following H. Busemann. Recall that $\Lambda(x,y,u,v)$ denotes the quotient of distances $d(x,y) / d(u,v)$, which is uniquely definable (as opposed to $d(x,y)$).

(521) <u>Axiom R8</u>:

There exists a neighbourhood U in P such that all points $u,v,x,y,z \in U$ satisfy: If

$\Lambda(x,u,u,z) = \Lambda(x,v,v,y) = 1$ and

$\Lambda(u,z,x,z) = \Lambda(v,y,x,y) = \frac{1}{2}$, then

$\Lambda(u,v,z,y) = \frac{1}{2}$.

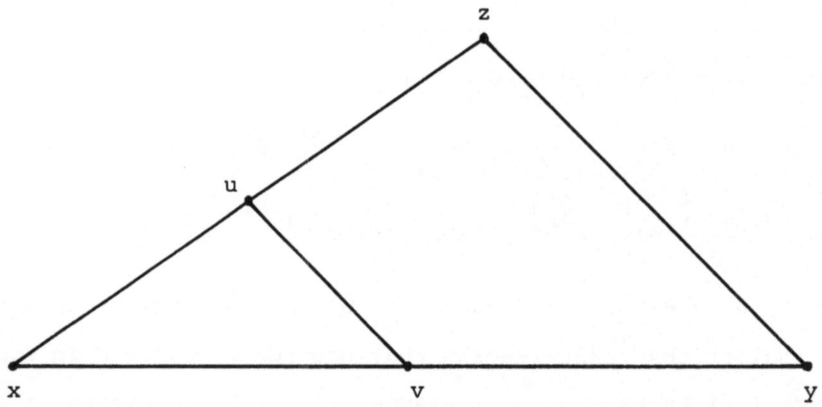

fig. (522)

From [BUS] § 41 it follows, that the curvature tensor vanishes identically.

5.3 EUCLIDEAN REPRESENTATION

In order to state the main theorem of this book it seems appropriate to define more precisely what is understood by the species of structure "Euclidean space" $\sum_3$. It is given by

(531) (i) principle base sets E, V,

(ii) an auxiliary base set $\mathbb{R}$,

(iii) a structural term

$$(+,\cdot,\langle|\rangle,o) \in P(V\times V\times V) \times P(\mathbb{R}\times V\times V) \times P(V\times V\times\mathbb{R}) \times P(V\times E\times E),$$

(iv) and axioms expressing that $(V,\mathbb{R},+,\cdot,\langle|\rangle)$ is a 3-dimensional, real, linear space equipped with an (positively definite, symmetric, bilinear) inner product $\langle|\rangle$, and its additive group $(V,+)$ operates transitively and freely on E (via o). We will set $\alpha = (E,V,\mathbb{R};+,\cdot,\langle|\rangle,o)$.

Often an "orientation" on E is considered as an additional component of the structural term, but this is not necessary in our context.

It is well-known that any two structures of this kind are $\sum_3$-isomorphic, i.e. $\sum_3$ is <u>categorical</u>. We will gather some widely used definitions.

(532) Definition: Let $\alpha = (E,V,\mathbb{R},+,\cdot,\langle|\rangle,o)$ be a structure of the kind $\sum_3$.

A mapping $T_v : E \to E$ of the form $A \mapsto v \circ A$, $v \in V$, is called a <u>translation</u>. A linear mapping $D : V \to V$ satisfying $\langle Dx|Dx\rangle = \langle x|x\rangle$ for all $x \in V$ and $\det D > 0$ is called a <u>proper rotation</u> of V and $\tilde{D} : E \to E$, defined by $\tilde{D}(v\circ A) = (Dv) \circ A$, for some fixed $A \in E$, is called a <u>proper</u> <u>rotation</u> of E with center A. The group generated by translations and proper rotations of

E will be denoted by $T(\alpha)$.

The linear mapping $S : V \to V$, $Sv \underset{def}{=} -v$, is called a (point) <u>reflection</u>; $\tilde{S} : E \to E$, defined by $S(v \circ A) = (Sv) \circ A$ for some fixed $A \in E$, is called a reflection of E with center A.

Analogously, the <u>dilatations</u> L_λ, $\lambda > 0$, are defined:

$L_\lambda : V \to V$, $L_\lambda v \underset{def}{=} \lambda v$; $\tilde{L}_\lambda : E \to E$, $\tilde{L}_\lambda(v \circ A) \underset{def}{=} (L_\lambda v) \circ A$.

Finally, $\tau(\alpha)$ will be defined as the class of all open subsets of E, $R(\alpha)$ as the subclass of open, bounded subsets.

Now let $\sum_{TF}$ (<u>T</u>its/<u>F</u>reudenthal) be the species of structure given by $(S;\tau,G)$ such that (S,τ) is a topological space and G a group of homeomorphisms and the further axioms of Freundenthal [FRE] are satisfied (at least in the slightly weaker version used in this book). Recall that the set of points P (327) is endowed with a topological structure N^{top} associated with the uniformity N (333).

A glance at Freudenthal's list shows at once:

(533) Theorem: Assume the axioms R1 to R8 holding for $(R,<,T)$. Then there exists a structure α of the kind $\sum_3$ ("Euclidean space") such that $(P,N^{top},\bar{T})$ is $\sum_{TF}$-isomorphic to $(E,\tau(\alpha),T(\alpha))$. Consequently, $(\bar{R},\subset,\bar{T})$ is $\sum_2$-isomorphic to $(R(\alpha),\subset,T(\alpha))$.

The situation can be graphically represented in the following manner:

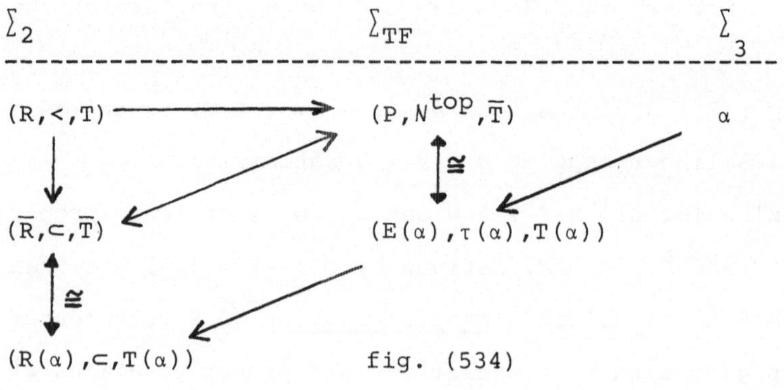

fig. (534)

The representation of $(\bar{R},\subset,\bar{T})$ by $(R(\alpha),\subset,T(\alpha))$ is essentially unique, since $\sum_{TF}$ is categorical. However, it will be instructive to compute the group Γ of all $\sum_{TF}$-automorphisms of a given Euclidean space $E(\alpha)$ equipped with the structure $\tau(\alpha)$, $T(\alpha)$. This group will be larger than $Aut_{\sum_3}(\alpha)$ corresponding to the fact that the Euclidean structure is only partially physically relevant. If $E(\alpha)$ is chosen as the space $\mathbb{R}^3$, Γ will be the class of admissible coordinate transformations. Therefore we may define:

(535) The group $\Gamma \underset{def}{=} Aut_{\sum_{TF}}(E(\alpha),\tau(\alpha),T(\alpha))$ will be called the <u>symmetry</u> <u>group</u> of physical geometry.

(536) Theorem: Γ is the group generated by translations, proper rotations, reflections and dilatations of $E(\alpha)$.

It will be adequate to give an elementary proof for this elementary result, not using the machinery of Lie algebra theory or projective geometry.

Proof: Let $\gamma \in \Gamma$. That is, $\gamma : E(\alpha) \to E(\alpha)$ will be a homeomorphism satisfying (*) $\gamma\, T(\alpha)\, \gamma^{-1} = T(\alpha)$.

If $A \in E$ and $\gamma A = v \circ A = T_v A$, then $\alpha \underset{def}{=} T_{-v}\, \gamma$ maps A onto A. Hence $\alpha\, J_A\, \alpha^{-1} = J_A$ if J_A is defined to be the group of proper rotations with center A.

More generally, $\alpha\, J_B\, \alpha^{-1} = J_C$, if $\alpha B = C$.

Each rotation $D \in J_A$ is a rotation around an axis $\vec{a}(D) \in V$ with some angle $\varphi(D) \in [0,2\pi[$. We claim:

$\varphi(\alpha D \alpha^{-1}) = \varphi(D)$.

Proof: Consider cyclic subgroups of J_A of order n, defined by $D^n = 1$, $\vec{a}(D) = \text{const}$. Under $D \to \alpha\, D\, \alpha^{-1}$ they are mapped 1 : 1 onto cyclic subgroups of the same order, hence their angle of rotation is conserved: $\varphi(D) = \frac{2\pi k}{n} = \varphi(\alpha D \alpha^{-1})$. Now, the union of all finite cyclic subgroups is dense in $J_{A,\vec{c}} \underset{def}{=} \{D \in J_A | \vec{a}(D) = \vec{c}\}$ and the claim follows by continuity of α.

Further, α throws the set of fixed points of any $D \in J_B$ onto the set of fixed points of $\alpha\ D\ \alpha^{-1}$ in an 1 : 1 fashion, and thus maps lines onto lines.

Especially, if we denote by $\check{\alpha}$ the lift of α to V, $\check{\alpha}$ permutes the lines through $O \in V$ conserving the angles between them.
The latter follows from the equivalence
$\measuredangle(\vec{a}(D_1), \vec{a}(D_2)) = \varepsilon \Leftrightarrow \exists\ R \in J_A$ such that $\varphi(R) = \varepsilon$ and $D_1 = R\ D_2\ R^{-1}$.

Now there exists a (linear) rotation β of V which induces the same permutation of the lines through O as $\check{\alpha}$, because the coefficients of a linear equation between unit vectors
$\vec{e}_3 = \lambda_1 \vec{e}_1 + \lambda_2 \vec{e}_2$
may be expressed as functions of the angles $\measuredangle(\vec{e}_1, \vec{e}_2)$, $\measuredangle(\vec{e}_2, \vec{e}_3)$, $\measuredangle(\vec{e}_1, \vec{e}_3)$.
This result may be written as $\check{\alpha}\ \check{D}\ \check{\alpha}^{-1} = \beta\ \check{D}\ \beta^{-1}$ for all proper rotations $\check{D}$ in V. Or, $\delta \underset{def}{=} \beta^{-1}\ \check{\alpha}$ commutes with all such $\check{D}$. Hence δ maps orbits of the group $\check{J}_A$ onto orbits, that is, spheres onto spheres, and leaves invariant the fixed point sets of $\check{D}$, that is, the lines through O. Further, together with α, δ maps lines onto lines.
Consider the 3 noncollinear points $O, x, y \in V$ and the plane P spanned by them. Clearly, δ leaves P invariant. Let the line g through x and y be mapped onto δg. Let us prove "g parallel δg."
If $z \in g \cap \delta g$, then $\delta^{-1} z$ and z would lie on g in contradiction to $O \notin g$ and the invariance of the line through O and $\delta^{-1} z$ under δ, if $\delta z \neq z$. If $\delta z = z$, δ is constant on the circle around O through z (by the properties of δ statet above and its continuity). If $g \neq \delta g$, one of these lines will intersect the circle at a second point w.

Then from $\delta z = z$ and $\delta w = w$ it follows that $\delta g = g$. Hence g is parallel to δg at any case.

Now, it follows by the familiar fact that parallel lines intercept proportional segments on transversals, that δ is a dilatation or a product of a dilatation and a reflection.

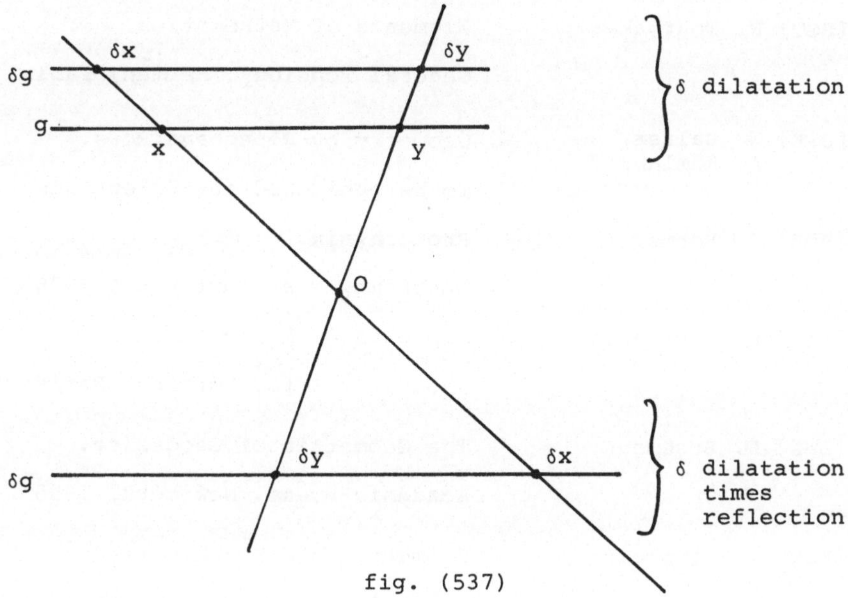

fig. (537)

The converse, that translations, rotations, dilatation and reflections satisfy (*), is immediate. □

6. REFERENCES

[ABR] R. Abraham: Foundations of Mechanics. Benjamin, New York, 1967

[B&C] F. Brickell, R.S. Clark: Differentiable Manifolds. Van Nostrand Reinhold Comp., New York, 1970

[BGT] N. Bourbake: Elements of Mathematics. General Topology. Herman, Paris, 1966

[B&K] W. Balzer, A. Kamlah: Geometry by Ropes and Rods. To be published in: Erkenntnis

[BÖH] G. Böhme: Protophysik. Suhrkamp, Frankfurt a.M., 1976

[BTS] N. Bourbaki: Elements of Mathematics. Theory of Sets, Hermann, Paris, 1968

[BUS] H. Busemann: The Geometrie of Geodesics. Academic Press, New York, 1955

[EIN] A. Einstein: Geometrie und Erfahrung. (27.1.1921) In: Mein Weltbild, Ullstein, Berlin, 1955

[FRE] H. Freudenthal: Neuere Fassungen des Riemann-Helmholtz-Lieschen Raumproblems. Math. Z. 63, (1955/56),374-405

[FRE 2] H. Freudenthal: Das Helmholtz-Liesche Raumproblem bei indefiniter Metrik. Math. Annalen 156 (1964), 263-312

[FRE 3] H. Freudenthal: Lie groups in the Foundations of Geometry. Adv. Math. 1 (1964), 145-19o

[F&V] H. Freudenthal, H. de Vries: Linear Lie Groups. Academic Press, New York, 1969

[GIL] R. Gilmore: Lie Groups, Lie Algebras, and Some of Their Applications. Wiley, New York, 1974

[HEL] H.v. Helmholtz: Über die Thatsachen, die der Geometrie zum Grunde liegen.
Nachr. Ges. Wiss. Göttingen 1868, 193-221.
Reprinted in: Über Geometrie, Wiss. Buchgesellschaft, Darmstadt, 1968.

[H&W] W. Hurewicz, H. Wallmann: Dimension Theory.
Revised Edition 1948, Princeton University Press

[K&N] S. Kobayashi, K. Nomizu: Foundations of Differential Geometry.
Wiley, New York, 1969

[LUD 1] G. Ludwig: Deutung des Begriffs "physikalische Theorie" und axiomatische Grundlegung der Hilbertraumstruktur der Quantenmechanik durch Hauptsätze des Messens.
Lecture Notes in Physics 4, Springer Berlin, 1970

[LUD 2] G. Ludwig: Einführung in die Grundlagen der Theoretischen Physik.
Band 1: Raum, Zeit, Mechanik.
Bertelsmann, Düsseldorf, 1974

[LUD 3] G. Ludwig: Grundstrukturen einer physikalischen Theorie.
Springer, Berlin, 1978

[MAY] D. Mayr: Zur konstruktiv-axiomatischen Charakterisierung der Riemann-Helmholtz-Lieschen Raumgeometrien und der Poincaré-Einstein-

Minkowskischen-Raumzeitgeometrien durch das Prinzip der Reproduzierbarkeit. Dissertation, München, 1979.

[MSZ] D.Montgomery, H. Samelson, L. Zippin: Singular points of a compact transformation group. Ann. of Math. (1) 63 (1959), 1-9

[REI] H. Reichenbach: Axiomatik der relativistischen Raum-Zeit-Lehre. In: A. Kamlah, M. Reichenbach (eds.) "Gesammelte Werke", Band 3, Vieweg, Wiesbaden, 1979.

[SCH 1] H.J. Schmidt: Zur Charakterisierung des Euklidischen Raumes durch Gebiete und Transporte. In: W. Balzer, A. Kamlah (eds.) "Aspekte der physikalischen Begriffsbildung", Vieweg, Wiesbaden, 1979.

[SCH 2] H.J. Schmidt: Entwurf einer Laborkinematik. Lecture held at the Conference "Grundstrukturen einer physikalischen Theorie: Raum, Zeit und Mechanik" in München, May 1979. To be published in the proceedings.

[TIT] J. Tits: Sur certaines classes d'espaces homogènes de groupes de Lie. Acad. Roy. Belg. Cl. Sci. Mem. Coll. 29 (1955), fasc. 268 pp.

[VAR] V.S.Varadarajan: Lie groups, and their Representations. Prentice-Hall, 1974.

[YAM] H. Yamabe: A generalization of a theorem of Gleason. Ann. of Math. (2) 58 (1953), 351-365

7. NOTATIONS

Most notations in this book are commonly used in set theory etc., perhaps with the exeption of the following:

$\mathcal{P}A$	the set of all subsets of A ,
$A = B \uplus C$	$A = B \cup C$ and $B \cap C = \emptyset$,
$\|M\|$	the number of elements of the finite set M,
$\sqsupset$	the inverse of the relation $\sqsubset$,
$\not\sqsubset$	the negation of the relation $\sqsubset$,
$f\|M$	the restriction of the map $f: A \to B$ to the subset $M \subset A$,
$[x]_\pi$ or simply $[x]^\pi$	the equivalence class of $x \in A$ w.r. to the equivalence relation π on the set A,
$[x]$	if x is a real number: the greatest integer less or equal x,
$[a,b]$	the closed,
$]a,b[$	open,
$[a,b[$	different
$]a,b]$	half-open intervals,
$a/b+c$	$\frac{a}{b} + c$.

Lecture Notes in Physics

Vol. 68: Y. V. Venkatesh, Energy Methods in Time-Varying System Stability and Instability Analyses. XII, 256 pages. 1977.

Vol. 69: K. Rohlfs, Lectures on Density Wave Theory. VI, 184 pages. 1977.

Vol. 70: Wave Propagation and Underwater Acoustics. Edited by J. Keller and J. Papadakis. VIII. 287 pages. 1977.

Vol. 71: Problems of Stellar Convection. Proceedings 1976. Edited by E. A. Spiegel and J. P. Zahn. VIII, 363 pages. 1977.

Vol. 72: Les instabilités hydrodynamiques en convection libre forcée et mixte. Edité par J. C. Legros et J. K. Platten. X, 202 pages. 1978.

Vol. 73: Invariant Wave Equations. Proceedings 1977. Edited by G. Velo and A. S. Wightman. VI, 416 pages. 1978.

Vol. 74: P. Collet and J.-P. Eckmann, A Renormalization Group Analysis of the Hierarchical Model in Statistical Mechanics. IV, 199 pages. 1978.

Vol. 75: Structure and Mechanisms of Turbulence I. Proceedings 1977. Edited by H. Fiedler. XX, 295 pages. 1978.

Vol. 76: Structure and Mechanisms of Turbulence II. Proceedings 1977. Edited by H. Fiedler. XX, 406 pages. 1978.

Vol. 77: Topics in Quantum Field Theory and Gauge Theories. Proceedings, Salamanca 1977. Edited by J. A. de Azcárraga. X, 378 pages 1978.

Vol. 78: Böhm, The Rigged Hilbert Space and Quantum Mechanics. IX, 70 pages. 1978.

Vol. 79: Group Theoretical Methods in Physics. Proceedings, 1977. Edited by P. Kramer and A. Rieckers. XVIII, 546 pages. 1978.

Vol. 80: Mathematical Problems in Theoretical Physics. Proceedings, 1977. Edited by G. Dell'Antonio, S. Doplicher and G. Jona-Lasinio. VI, 438 pages. 1978.

Vol. 81: MacGregor, The Nature of the Elementary Particle. XXII, 482 pages. 1978.

Vol. 82: Few Body Systems and Nuclear Forces I. Proceedings, 1978. Edited by H. Zingl, M. Haftel and H. Zankel. XIX, 442 pages. 1978.

Vol. 83: Experimental Methods in Heavy Ion Physics. Edited by K. Bethge. V, 251 pages. 1978.

Vol. 84: Stochastic Processes in Nonequilibrium Systems, Proceedings, 1978. Edited by L. Garrido, P. Seglar and P. J. Shepherd. XI, 355 pages. 1978

Vol. 85: Applied Inverse Problems. Edited by P. C. Sabatier. V, 425 pages. 1978.

Vol. 86: Few Body Systems and Electromagnetic Interaction. Proceedings 1978. Edited by C. Ciofi degli Atti and E. De Sanctis. VI, 352 pages. 1978.

Vol. 87: Few Body Systems and Nuclear Forces II, Proceedings, 1978. Edited by H. Zingl, M. Haftel, and H. Zankel. X, 545 pages. 1978.

Vol. 88: K. Hutter and A. A. F. van de Ven, Field Matter Interactions in Thermoelastic Solids. VIII, 231 pages. 1978.

Vol. 89: Microscopic Optical Potentials, Proceedings, 1978. Edited by H. V. von Geramb. XI, 481 pages. 1979.

Vol. 90: Sixth International Conference on Numerical Methods in Fluid Dynamics. Proceedings, 1978. Edited by H. Cabannes, M. Holt and V. Rusanov. VIII, 620 pages. 1979.

Vol. 91: Computing Methods in Applied Sciences and Engineering, 1977, II. Proceedings, 1977. Edited by R. Glowinski and J. L. Lions. VI, 359 pages. 1979.

Vol. 92: Nuclear Interactions. Proceedings, 1978. Edited by B. A. Robson. XXIV, 507 pages. 1979.

Vol. 93: Stochastic Behavior in Classical and Quantum Hamiltonian Systems. Proceedings, 1977. Edited by G. Casati and J. Ford. VI, 375 pages. 1979.

Vol. 94: Group Theoretical Methods in Physics. Proceedings, 1978. Edited by W. Beiglböck, A. Böhm and E. Takasugi. XIII, 540 pages. 1979.

Vol. 95: Quasi One-Dimensional Conductors I. Proceedings, 1978. Edited by S. Barišić, A. Bjeliš, J. R. Cooper and B. Leontič. X, 371 pages. 1979.

Vol. 96: Quasi One-Dimensional Conductors II. Proceedings 1978. Edited by S. Barišić, A. Bjeliš, J. R. Cooper and B. Leontić. XII, 461 pages. 1979.

Vol. 97: Hughston, Twistors and Particles. VIII, 153 pages. 1979.

Vol. 98: Nonlinear Problems in Theoretical Physics. Proceedings, 1978. Edited by A. F. Rañada. X, 216 pages. 1979.

Vol. 99: M. Drieschner, Voraussage – Wahrscheinlichkeit – Objekt. XI, 308 Seiten. 1979.

Vol. 101: A. Martin-Löf, Statistical Mechanics and the Foundations of Thermodynamics. V, 120 pages. 1979.

Vol. 102: H. Hora, Nonlinear Plasma Dynamics at Laser Irradiation. VIII, 242 pages. 1979.

Vol. 103: P. A. Martin, Modèles en Mécanique Statistique des Processus Irréversibles. IV, 134 pages. 1979.

Vol. 104: Dynamical Critical Phenomena and Related Topics. Proceedings, 1979. Edited by Ch. P. Enz. XII, 390 pages. 1979.

Vol. 105: Dynamics and Instability of Fluid Interfaces. Proceedings, 1978. Edited by T. S. Sørensen. V, 315 pages. 1979.

Vol. 106: Feynman Path Integrals, Proceedings, 1978. Edited by S. Albeverio et al. XI, 451 pages. 1979.

Vol. 107: J. Kijowski, W. M. Tulczyjew, A Symplectic Framework for Field Theories. IV, 257 pages. 1979.

Vol. 108: Nuclear Physics with Electromagnetic Interactions. Proceedings, 1979. Edited by H. Arenhövel and D. Drechsel. IX, 509 pages. 1979.

Vol. 109: Physics of the Expanding Universe. Proceedings, 1978. Edited by M. Demiański. V, 210 pages. 1979.

Vol. 110: D. A. Park, Classical Dynamics and Its Quantum Analogues. VIII, 339 pages. 1979.

Vol. 111: H.-J. Schmidt, Axiomatic Characterization of Physical Geometry. V, 163 pages. 1979.